T0337007

Insight on Environmental Genomics

Series Editor
Françoise Gaill

Insight on Environmental Genomics

The High-throughput Sequencing Revolution

Denis Faure
Dominique Joly

ELSEVIER

First published 2016 in Great Britain and the United States by ISTE Press Ltd and Elsevier Ltd

Apart from any fair dealing for the purposes of research or private study, or criticism or review, as permitted under the Copyright, Designs and Patents Act 1988, this publication may only be reproduced, stored or transmitted, in any form or by any means, with the prior permission in writing of the publishers, or in the case of reprographic reproduction in accordance with the terms and licenses issued by the CLA. Enquiries concerning reproduction outside these terms should be sent to the publishers at the undermentioned address:

ISTE Press Ltd
27-37 St George's Road
London SW19 4EU
UK

Elsevier Ltd
The Boulevard, Langford Lane
Kidlington, Oxford, OX5 1GB
UK

www.iste.co.uk

www.elsevier.com

Notices
Knowledge and best practice in this field are constantly changing. As new research and experience broaden our understanding, changes in research methods, professional practices, or medical treatment may become necessary.

Practitioners and researchers must always rely on their own experience and knowledge in evaluating and using any information, methods, compounds, or experiments described herein. In using such information or methods they should be mindful of their own safety and the safety of others, including parties for whom they have a professional responsibility.

To the fullest extent of the law, neither the Publisher nor the authors, contributors, or editors, assume any liability for any injury and/or damage to persons or property as a matter of products liability, negligence or otherwise, or from any use or operation of any methods, products, instructions, or ideas contained in the material herein.

For information on all our publications visit our website at http://store.elsevier.com/

© ISTE Press Ltd 2016
The rights of Denis Faure and Dominique Joly to be identified as the authors of this work have been asserted by them in accordance with the Copyright, Designs and Patents Act 1988.

British Library Cataloguing-in-Publication Data
A CIP record for this book is available from the British Library
Library of Congress Cataloging in Publication Data
A catalog record for this book is available from the Library of Congress
ISBN 978-1-78548-146-8

Printed and bound in the UK and US

Contents

Preface

Environmental genomics brings together the various fields of knowledge on past and present living organisms and ecosystems, through the analysis of the sequence of genes and genomes using (meta)barcode or (meta)genomics as well as the analysis of genes and genomes expression using (meta)transcriptomics. Combined with other approaches such as (meta)proteomics, metabolomics and *in situ* observations, environmental genomics provides a wealth of information about the taxonomy and diversity of current and fossil organisms, their phylogeny and evolution, their potentialities and abilities to adapt and acclimatize, their biology, their functional traits and their interactions with the environment in its biotic and abiotic dimensions.

Next-generation DNA and RNA sequencing technologies (NGS) allow the production of structural and functional (meta)genomics data at a scale that was unimaginable just a few years ago. NGS technologies have significantly and permanently changed, not only the experimental strategies used to investigate evolution, biodiversity and the ecology of living and dead organisms and ecosystems, but also the scientific perspective on human populations. They have also profoundly changed our representation of living organisms and their intrinsic characteristics and interactions with other organisms. NGS technologies apply to all living organisms (Archaea, Eukaryotes and Bacteria and their associated viruses) and give access to previously unknown taxonomic groups and biological functions. This technological revolution opens up new scientific opportunities as it renews technical and methodological pathways. Environmental genomics is the field that benefited and still benefits the most from this paradigm shift, as it integrates the most advanced knowledge of environmental science at all scales, whether these are biological (from the metabolite to ecosystem), spatial (from the local to the planetary level), or temporal (from the possibly very distant past to recent or current time). Harnessing these new possibilities requires specific efforts in terms of the initial and continued education of the various members of the scientific communities (biology, ecology, but also computer science and mathematics), especially those who will conduct research in the years to come.

This book is the result of a collective effort of reflection and writing of the French scientific community, involved in the research group on environmental genomics, the successor of the Réseau Thématique Pluridisciplinaire of the same name (http://gdr3692.wix.com/gdrge). Its intent is to present the possibilities of this emerging discipline (Introduction), to provide an overview of the issues, challenges and perspectives of environmental genomics (Chapter 1) and to present the technological revolutions brought about by the development of the new sequencing methods, identifying their possibilities and limitations (Chapter 2). The following chapters describe in more detail the state of thinking in various methodological and scientific fields: access to NGS data production and sharing (Chapter 3), NGS data quality (Chapter 4), characterization of life (Chapter 5), structure and dynamics of biodiversity (Chapter 6), study of the evolution and adaptation of genes and genomes (Chapter 7), analysis of degraded and/or ancient DNA (Chapter 8), functional and genomic ecology of populations (Chapter 9) and communities (Chapter 10). The last chapters discuss more exploratory aspects of the modeling and functioning of ecosystems (Chapter 11) as well as aspects related to the future developments of 3^d and 4th generation technologies (Chapter 12).

Let us wager that the new research possibilities enabled by environmental genomics, combined with new methods for the development and analysis of complex systems, will enable the coming generations of students and researchers to understand all dimensions of life, taking into account the complexity of the intricate interactions that occur amongst organisms, their functional relations and beyond their complex interactions with the environment which they shape and which shapes them. The reader will discover throughout the pages of this book how interdisciplinarity is at the core of this emerging research domain, as it requires skills and expertise that pertain to various fields of biology (ecology, evolution, paleobiology, taxonomy, etc.), of "hard" sciences (computer science and bioinformatics, mathematics, biogeochemistry, physics, etc.) as well as of human and social sciences (sociology, anthropology and paleoanthropology, statistics, etc.). Reviving the naturalist approaches of the 18th century, environmental genomics is a field science that today offers an exceptional opportunity to tackle the societal issues raised by global change of natural or antrhopic origin.

Denis FAURE
Dominique JOLY
April 2016

Acknowledgments

We wish to express our deep gratitude to Françoise Gaill (CNRS, France), who encouraged us to propose this book, and Stéphanie Thiébaut (CNRS, France), who supported us during the development of this work.

This book grew from the involvement of many contributors who helped us to develop and write the chapters in which they shared their knowledge and their most recent discoveries, as well as their vision and prospective ideas for this emerging research field, environmental genomics. We especially wish to thank Sylvie Salamitou, who coordinated the relations with the authors and with the French national photo libraries, as well as Philippe Bertin, Damien Eveillard, Catherine Hänni, Mathieu Joron, Line Le Gall, Guillaume Lecointre, Francis Martin, Eric Pelletier, Guy Perrière, Pierre Peyret, François Pompanon, Xavier Raynaud, Sarah Samadi, Télesphore Sime-Ngando and Xavier Vekemans for having agreed to coordinate some chapters.

The imagery of this work comes from various contributions and sources. We particularly thank the researchers who provided the pictures and schemas, the managers of the CNRS image library well as Patrice Vagnon and Pierre Ferrière who made photographs of their personal collections available to us.

Last but not least, this book would not have become a reality without the support of the publishing and editing teams of ISTE Ltd; we sincerely thank them.

Introduction

Environmental genomics brings together the various fields of knowledge on past and present organisms and ecosystems, through the analysis of the nucleotide sequence* of genes, genomes, metagenomes, transcripts, transcriptomes and metatranscriptomes. Combined with other technologies and observations, environmental genomics provides a wealth of information on the taxonomy* and diversity of current and fossil organisms, their phylogeny* and evolution, their potential and ability to adapt and acclimatize, their biology, their functional traits and their interaction with the environment in its biotic and abiotic dimensions.

Figure I.1. *Environmental DNA sequencing enables the reconstitution of visible or invisible organism communities in an environment,* © *Pierre Ferrière*

The first DNA sequencing technologies emerged in the seventies and were used to characterize the genes of organisms as well as to detect them in their environment. The first occurrence of the term "environmental DNA" dates back to a scientific publication of [OGR 87]. As soon as the end of the 1990s, the sequencing of the first genomes and metagenomes was achieved. The first sequenced genome was that of the bacterium *Haemophilus influenzae* (1.8 Mb) in [FLE 95]. The first generation of sequencing technologies contributed to the emergence of structural and functional genetics and genomics, of phylogenetics and molecular taxonomy* and to the development of molecular markers of taxons or barcodes. The second generation of DNA sequencing technologies emerged in the middle of the first decade of the 21st century, with LifeTechnologies' SOliD sequencers, Illumina's* Solexa and Roche's 454 pyrosequencing*.

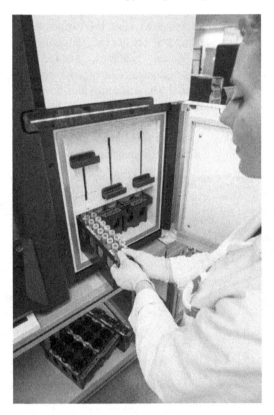

Figure I.2. *Installation of reactives in an Illumina HiSeq 2000 fast-rate sequencer, © Cyril Fresillon/CNRS Photothèque CNRS 20130001_1487*

A substantial improvement of Illumina's technology occurred in 2007, enabling both the production of larger numbers of DNA sequences and drastically reduced costs. Thanks to the conjunction of these two major improvements, NGS technologies became rapidly widespread in public and private sector research laboratories involved in health or environmental sciences. The third generation of sequencers is emerging: these sequencers enable researchers to read the DNA directly, bypassing the amplification stage, as for example in the approach of Pacific Biosciences (PacBio). Nowadays, incredible improvements of sequencers, in terms of data generation rate and accuracy, as well as improvements in computational techniques for the analysis of DNA sequences, are so significant that NGS technologies have become accessible to smaller research teams as well as larger research centers. Technologies now go to miniaturisation through the pocket-sized MinIon (Oxford TEchnol.), able to deliver sequencing data by adding the sample to the flow cell and after plugging the device into a PC or laptop using a USB.

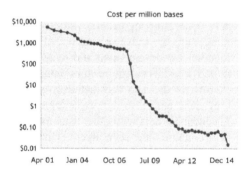

Figure I.3. *Significant drop in sequencing costs over recent years. From 2001 to 2007, sequencing costs, expressed in USD per million bases (Mb) are estimated on the basis of the use of first-generation technology and its improvements (Sanger). From January 2008, costs are estimated on the basis of second-generation methods (Illumina). Updated data is available from the National Human Genome Research Institute (USA) (www.genome.gov/ sequencingcosts), © Denis Faure*

These recent technological developments enable the study of DNA and RNA with unprecedented power, whether the nucleic acids are extracted from a single cell, from populations or even from communities of individuals. NGS technologies offer new scientific opportunities for the science of environment, evolution, ecology and biodiversity. The development of this know-how is to be understood within the context of intense international competition to produce knowledge, but also within the context of societal issues such as biodiversity or the evaluation of the impact of human activities on the environment. NGS also offers the opportunity for individual researchers and laboratories to share and exchange larger data sets and to combine their expertise around common objects to study more effectively. From another perspective, the combination of NGS with biological and medical data about human populations raises ethical and legal issues.

The new opportunities offered by NGS have a cost, as these technologies require the acquisition and harnessing of new tools for the analysis, processing, storage and exchange of NGS-generated data. Therefore, new challenges have emerged that were overlooked during the emergence of NGS. They mainly involve the processing of the tremendous amounts of data produced by NGS: DNA sequencers can identify over one million DNA bases per second, i.e. more than one human genome (3, 400 Mb) per hour. The accuracy of the primary data is also a major issue, in order to avoid any bias or interpretation error and to obtain robust statistical significance thresholds, especially when studying genetic variations. The development of tools for the online analysis and exploitation of data also sets major challenges. For example, the development and long-term code maintenance of online services and interfaces require substantial software engineering work. Setting up norms for the data and metadata is also necessary for the sequences stored in data banks to be exploitable. Similarly, issues of interoperability of sequencing data with other data and metadata types whose origins, formats and structures are very diverse (for example, measures of environmental parameters) are also of major importance. Lastly, appropriately sized computer infrastructures are necessary to enable the safe storage and processing of these vast amounts of data.

Figure I.4. *Data center, © BalticServers_data_center, Wiki commons*

This colloborative book proposes to discover the opportunities that environmental genomics opens to environmental science (Chapter 1), to understand the technical and bioinformatics changes that are related to the production, processing and accuracy of NGS-generated data (Chapters 2, 3 and 4), and to discover the new NGS-based approaches that are developed in major fields of environmental science: biodiversity (Chapters 5 and 6), the evolution and adaptation of organisms in past and present ecosystems (Chapters 7 and 8), the ecology of populations and communities of organisms (Chapters 9 and 10) and the study of the dynamics of ecosystems (Chapter 11). The last chapter outlines the prospects for developments of new approaches and questions in environmental science (Chapter 12).

The book will refer to examples of environmental genomics research from recently published books and articles. These examples illustrate the extent of fields of application and scientific issues that arise from this rapidly expanding research domain.

Where terms are marked with the * symbol, please refer to the Glossary for general and technical explanations.

Figure I.5. *Computer-generated image of a diatom, or* Phaeodactylum tricornutum. *It is the second marine diatom whose genome has been sequenced [BOW 08]. It is polymorphous with 3 morphotypes (3 shapes are visible here),* © *Chris Bowler/Alessandra de Martino/David E Gilbert/CNRS Photothèque CNRS 20090001_0203*

1

Issues, Challenges, Scientific Bottlenecks and Perspectives

Environmental genomics benefits from the extraordinary development of NGS technologies, which redefine what research can do in the fields of ecology, evolution and environmental science. It is now possible to read the still widely unknown fraction of biodiversity, of which 80 to 90% are not taxonomically described [MOR 11]. The taxonomic description of organisms as well as the understanding of their functioning and evolution has become the major endeavor of 21st century biology [LOS 13]. To meet this challenge, academic researchers are getting involved in approaches that are named *omics*, thus erasing the boundaries between the various disciplines. Research progresses thanks to a sustained back-and-forth exchange between field work and laboratory work, integrating environmental data and metadata that was gathered at micro- and macroscopic scales (from cell to ecosystem). Environmental genomics is part of this paradigm shift, as it favors the interdisciplinary mixing of points of view on the complexity and functions of genes and organisms, as well as on their dynamics and evolution through space and time.

NGS technologies enable more exhaustive access than ever to describe biodiversity (*DNA barcode, metagenomics, molecular phylogenetics*) and to understand its dynamics and evolution (*evolutionary genomics, population genetics, molecular ecology, ecology of communities*). A major stake in this new field is the characterization of biodiversity, from the individual to the population, from the populations to the communities and from the communities to the ecosystem [JOL 15]. Another crucial challenge is to understand the dynamics of this biodiversity in order to be able to estimate environmental impacts, whether they are of anthropic or natural origin.

Figure 1.1. *Coral reef, © Erwan Amice/IRD/CNRS/*
CNRS Photothèque 20110001_0799

Most of the data that exists today is restricted to a few specific organisms: in the animal kingdom, some mammals, birds and fish, some species of trees and flower plants as well as some alga, fungi, bacteria and viruses. Whole sections of the tree of life remain unknown. For example, estimates suggest that one gram of soil contains more than 50,000 species of microbes [ROE 07]. We are still very far from having an exhaustive image of the taxonomic richness and distribution of genomes of macroscopic (insects, crustaceans, fungi) and microscopic organisms, be they autonomous, parasites or symbiotic (viruses, protozoa, bacteria or helminth; the latter representing 50% of the specific diversity). A major effort remains to be made to coordinate and diversify ecological data collections in order to better document the specific marine and terrestrial richness of our planet (see Chapter 4), understand its evolution (especially within the context of global change, whether it is of climate or human origin, see Chapters 7 and 8) and to propose actions to protect or manage the environment, taking into account scenarios that predict the evolution of biodiversity (see Chapter 11).

Figure 1.2. *Field collection of soil from a parcel of the Great Green Wall for microbiological analysis, © Axel Ducourneau/ CNRS Photothèque 20110001_1233*

That these high throughput technologies gain momentum so rapidly makes it necessary to build shorter and more controlled circuits in order to enable the long term preservation of biological resources, their analysis, the setting up of databases (linked to associated metadata and functional correlations [HAR 13]) and eventually the exploitation of results. All the complex systems of the assembly of living organisms must be involved, even those that represent natural ecosystems and agro-ecosystems, in order for a complete spatio-temporal record to be available. This is necessary to enable the retrospective analysis of the evolution of various segments of the environment. Such a challenge can only be met with a multi- and interdisciplinary approach from disciplines as diverse as biology, bio-computing, physics and mathematics, as well as human and social sciences.

1.1. Environmental DNA

The study of environmental DNA gained significant momentum during the first decade of the 21st century, mostly within communities of microbiologists who studied samples of soil, air or water [TAB 12]. The first methodological revolution that this approach propelled consists of extracting DNA without isolating the various organisms it pertains to. Environmental DNA (eDNA) is, thus, a mix of possibly degraded DNA of the various organisms that exist in the local environment. It consists of cellular DNA from the living or dead cells and of extracellular DNA released by the cells. This approach solves two difficulties, as it gives access to the DNA of organisms that could not be grown in the laboratory, as well as to ancient DNA (aDNA) contained in the tissues of dead organisms.

The analysis of environmental DNA with NGS can be achieved either by selective amplification followed by the sequencing of genes of interest, in a metabarcode* approach, or by the sequencing of the whole of the environmental DNA, in a metagenomic approach.

Figure 1.3. *Sample collection. From left to right, litter in Guyana forest (Nouragues), © Claude Delhaye/CNRS Photothèque 20090001_1577; coral reef in Crete, © Thierry Perez/CNRS Photothèque 20140001_0367; sand storm and its flock of microorganisms, © Françoise Guichard/Laurent Kergoat/CNRS Photothèque 20050001_0870*

The metabarcode approach enables the analysis of the diversity of organism communities of current and past times. It uses genetic markers of the barcode type, which are specific to a particular taxon (depending on scale, a taxon is specific to the species, the genus or the family) such as the *16S rDNA* for bacteria, the *18S rDNA*, *Cyt-b* or *COI* for animals, *rbcL* and *matK* for plants or *ITS* for fungi. This approach can also be used to determine the biodiversity of genes that code for ecologically interesting functions, like those involved in the nitrogen cycle in soil bacteria.

Analyzing environmental DNA using a metagenomic approach enables us to apprehend three important aspects in community ecology:

1) biodiversity in terms of taxonomic richness;

2) identification of the functions that are coded by the genes of organisms;

3) the assembly of whole genomes, including the genomes of microorganisms that can't be cultivated in laboratory conditions or simply of ancient organisms that are extinct today.

Figure 1.4. *Wasp trapped in Miocene amber, © Vincent Perrichot/CNRS Photothèque 20120001_1756*

1.2. Consortia and international networks

Large international comparative genomics programs that study thousands of species of a particular taxonomic group all share common aims; to document biological diversity, to define phylogenic relationships, to identify possible links between functional traits and genetic patterns, and potentially contribute to our understanding of the biological processes that have significant economic or societal impacts. These include the carbon cycle, metabolic networks, the fight against pathogens, emerging infectious diseases, Darwinian medicine and the resilience of organisms and ecosystems to global change. This section briefly presents the main international consortia, whether they focus on biological models or specific ecosystems.

Figure 1.5. *Molluscs collected during the Tara Oceans expedition. Today, biological expeditions collect a previously unsuspected biological diversity, thanks to the current methods of investigation. To relate a morphological specimen to its genomic description is a real challenge for modern taxonomy. © Christian Sardet/Tara Océans/CNRS Photothèque 20100001_0972*

1.2.1. Fungi

The programs *1,000 Fungal Genomes* (1KFG) and *Mycorrhizal Genomics Initiative* focus on identifying the links between functional features and genetic patterns in various groups of fungi (saprophytes, symbionts or pathogens) that control the carbon cycle in terrestrial and aquatic ecosystems. These programs are run by the Joint Genome Institute (JGI) of the United States Department of Energy. Within the framework of the *1000 Fungal Genomes project* (Figure 1.6), the genome of two different species will be sequenced for each of the 500 known fungal families. This significant sequencing effort enables the exploration of the genomic diversity of these species and the identification of molecular mechanisms that determine their evolution and functional features. The project also involves the science of energy, environment and biodiversity. They target fungal genomes of three domains: vegetable health, biorefinery and fungal diversity. The sequencing and analysis of the genome of about 50 ectomycorrhizal symbionts, 100 pathogenic agents and 100 brown and white rots have already been completed. Comparing the inventories of genes of these fungi highlighted the key role they have in the coding of enzymes that decompose the polysaccharides of the plant cell wall (CAZyme) [FLO 12]. Molecular clock analyses suggest that the origin of lignin degradation might have coincided with the sharp decrease in the rate of organic carbon burial around the end of the Carboniferous period.

These studies also enabled the establishment of scenarios that describe the evolution of forest fungi: the ancestral species was probably a white rot that played a decisive part in the massive decomposition of the lignocellulose of the first trees colonising the continents, 300 million years ago; this functional group then diversified, eventually giving birth to brown rots and litter decomposers. Ectomycorrhizal fungi would originate from these functional groups, after the massive loss of their CAZymes and acquiring of protein effectors controlling immunity and the root development of the host plant.

Figure 1.6. Laccaria amethystina *is a symbiotic ectomycorrhizal fungus whose genome was sequenced within the framework of the Mycorrhizal Genomics Initiative program (http://mycor.nancy.inra.fr/IMGC/MycoGenomes/),* © *Francis Martin*

1.2.2. *Arthropods*

The *i5k* program was developed in 2011 on the basis of past experience sequencing the genomes of several species of arthropods, such as drosophila, bees, heteroptera, plant lice, beetles and other insects, as well as centipedes, scorpions, spiders, velt worms, and the other bilateria. It aims to sequence 5,000 genomes in 5 years, targeting arthropods of agricultural, food or medical interest as well as genomes that are potentially usable in energy production. This project complements the *Genome 10k* project of sequencing 10,000 genomes of vertebrates to balance the research effort more equally on the main branches of the tree of life. It addresses issues related to the fight against pests and pathogens, to forensic applications, aquaculture and biodiversity conservation. Its aims include deepening our knowledge of chemical-sensory reception mechanisms that serve as natural defense detectors, elucidating trophic relations between plants, insects and vectors as well as understanding the role of insects in carbon and methane capture, especially in the case of ants and termites.

Figure 1.7. Papilio machaon, *one of the species of the i5K program,* © *Patrice Vagnon*

1.2.3. *Vertebrates*

As previously mentioned, the *Genome 10k* project proposes a collection of genomic banks of tissues and DNA of 10,000 species of vertebrates, with the aim to gather one specimen of each known genus. This figure is currently a reasonable compromise between what technologies, current as well as those expected within the next few years, are able to provide, and a relevant coverage that would represent biodiversity. The program aims to document the genetic modifications, including reorganization, duplications as well as gains and loss of genes that have shaped the diversity of past and current forms as well as to reconstruct their evolutionary history. This knowledge, as a whole, will constitute an essential reference resource for proposals of new paradigms in evolutionary biology and for the understanding of functions that are essential to life.

1.2.4. Human societies – health

NGS provide the study of our species with a huge body of data. Within the framework of the *1,000 Genomes* sequencing project, this data enables us to document the genetic variations whose frequency is less than 1% when different human populations are compared. Within the framework of the *ENCODE* program, devoted to the characterization of functional elements of the human genome, the collected data can be used to reconstruct the history of the human lineage, but also to fight rare or common diseases that arise from our industrial societies, such as some cancers [GAT 11, AIM 13, JOB 13, SOU 14]. Various international projects, driven for example by the *US National Institute of Health* (NIH), address issues related to human health via institutes such as the *National Institute of Environmental Health Sciences* (NIEHS) devoted to human-environment interactions or the *National Human Genome Research Institute* (NHGRI) interested in human biology. The former aims to document the impact of the environment on human health, via 11 lines of work devoted to the study of the biological mechanisms which control the response to environmental stress, with the end purpose of guiding the decision-makers in their policies. The latter (NHGRI) coordinates the Human Genome Project with a specific emphasis on ethical, legal and societal implications of advances in this field. In a partnership with the NIH, 7 countries take part in the *International Human Microbiome Consortium* (IHMC) project, whose aim is to set up a common strategy for the study of the human microbiome, to eventually improve disease prevention and cure. France has a historical leading role in this project, in particular with two flagship projects, *MetaGenoPolis*, which focuses on the human intestinal microbiome and is the continuation of the *MetaHIT* project. The *MetaHIT* project has notably shown that individuals fall into three different groups according to the microbes in their intestines and this, independently from their geographical origin, state of health (overweight or inflammation of the alimentary canal), sex or age. This classification, like that of blood groups, is individual-specific, leading researchers to define "enterotypes". The *MetaHIT* project also described the bacterial communities of two chronic diseases (obesity and inflammatory affection of the intestine) and developed a pilot project of functional metagenomic screening.

All these programs situate human individuals in their environment. The objectives are to better elucidate not only intrinsic human complexity, but also the diversity of physiological and molecular interactions each individual displays with the microorganisms it contains (microbiota) or is in contact with (environment). In this context, the role of environmental or ecological factors in the etiology of some diseases is considered today in an approach called "ecology of health". It consists of developing integrated research on public health, considering the ecological context in which human populations live. Within this framework, the emerging concept of "evolutionary medicine" [THO 13] replaces each individual, within a historical perspective, as the product of an evolution influenced by natural selection, with effects that differ according to climate, living conditions, physiological state, etc. This approach aims to develop individualized health care that takes into account the differences between individuals and between human populations.

Figure 1.8. *Fishermen in Trois-Sauts, Guyana, illustrating a project on the ecology of the bacterial resistance to* Escherichia coli *and* Staphylococcus aureus *in commensal flora (bacteria that live on a host without causing adverse effects) of humans and animals in a natural environment. The transition towards virulence and the impact on human health is also studied,* © François Catzeflis/CNRS Photothèque 20070001_0370

Beyond the considerable research effort focused on our species, for which NGS technologies renew our methods and knowledge, other programs also involve a wide international community. They include programs in fields for which research means have remained essentially descriptive until now. A non-exhaustive overview of these various programs follows below.

1.2.5. Microbiotes

The Earth Microbiome Project is a multidisciplinary project on a planetary scale, focused on microbial communities from a taxonomic and functional perspective. The global principle is to study microbial communities of each ecosystem according to common norms that apply to all and to collect environmental parameters of various ecosystems that represent diversity on Earth. More than 200,000 samples will be analyzed with methods pertaining to metagenomics, metatranscriptomics and metabarcode sequencing. In the long run, the purposes are to build a global Atlas, not only of genes but also of the proteins these genes encode, to build metabolic models of microbial communities that are specific to each environment and to achieve the assembly of about 500,000 microbial genomes. All this data will be available on an online portal enabling the visualization of the information. The success of such an endeavor relies on voluntary participation of researchers, provided they conform and contribute to the norms established by the *Earth Microbiome Project*.

1.2.6. Soils

The *Terragenome* project aims at the full sequencing of the metagenome of various soils (Figure 1.9). The project started with the study of the soil of the agricultural experimental station in Rothamsted. Recently, *Terragenome* extended its aims to other soils. It now integrates the works of various laboratories. The study of hundreds of thousands of soil microorganism species is driven by a metagenomic approach. Such a project has a great scale and requires a multidisciplinary approach associating scientists with complementary expertise (microbiology, microbial ecology, molecular biology, bioinformatics and soil physico-chemistry). The *Terragenome* consortium organizes yearly symposiums that enable the exchange and harmonization of the various experimental and bioinformatics methodologies, with the financial support of national agencies such as the Thuinen Institute in Germany or the NSF in the United States.

Figure 1.9. *The variety of landscapes and their soils is scrutinized in the Terragenome project, © Patrice Vagnon*

1.2.7. Marine genomics

The *Oceanomics* (wOrld oCEAN bioResources, biotechnologies and Earth-systeM serviCeS) project is both a fundamental and applied research project. Over a seven-year period, it groups 10 academic partners, six private partners and many other partners that are not directly financed but wish to collaborate. The *Oceanomics* project aims to understand the complexity of biological systems as well as the biotechnological potential of the greatest planetary ecosystem: oceanic plankton. It relies on the thousands of eco-morpho-genetic samples and data collected during the *Tara-Oceans* expedition [BOR 15]. *Oceanomics* will first of all explore this unique collection that covers all the planktonic communities, including viruses. A combination of NGS sequencing and fast-rate imaging protocols has been set up to extract taxonomic information from these biological samples. Comparing this data to environmental metadata* will lead to a deep taxonomic, metabolic and ecosystemic understanding of the structure, dynamics and evolution of planktonic diversity. This ecosystemic approach raises considerable hopes. Plankton accounts for 98% of the volume of our biosphere: we can only imagine the huge potential resource of still unknown lifeforms and unexplored bioactive compounds. Once the knowledge of this biodiversity is consolidated, the *Oceanomics* project will turn to collaboration with its private partners in order to:

1) transfer the new sequencing and fast-rate imaging technologies and methods to aquatic biomonitoring* case studies;

2) realize the phenotyping of environmental samples and chosen strains to analyze their lipids, secondary metabolites, exometabolomes (metabolomes excreted into the extracellular environment);

3) screen the chosen strains for their bioactive compounds that present some pharmaceutical or nutritional interest (in terms of beneficial effects on health) or some interest for aquaculture, cosmetics or any agricultural or environmental field. In parallel to its scientific activities, *Oceanomics* will serve as a study case to define a balanced legal model for the bioprospection of marine plankton, still mostly unused beyond national borders and consequently lying at the extreme limits of the current regulatory frameworks.

Figure 1.10. *Example of morphologic biodiversity observed in marine protists, © Colomban de Vargas*

1.2.8. Marine biotechnologies

The *Idealg – Seaweed for the future* project aims to develop the value of marine vegetables in a sustainable development framework. Marine macro-algae exhibit a great diversity in terms of phylogenies, ways of life, lifecycles, metabolic and cellular components they synthesize as well as genetic characteristics. They constitute a vast reserve for the development of new products and processes. The project focuses on the study of their genomics and post-genomics, to develop new tools and methods that enable the identification and selection of "resource" local populations that present industrially interesting properties. The development of new genetic tools (SNP*, QTL* and RAD* markers) and in some cases the elaboration of genetic maps of algae aim to better understand the fundamental mechanisms of adaptation and phenotype/environment interaction as well as to improve the methods of domestication of the populations. The final objective of this project is to contribute to the biotechnological development of the seaweed industry. The knowledge acquired will serve the development of the seaweed transformation industry (degradation, bioconversion, natural defenses, etc.). One of the aims of the project is to develop a "seaweedomics" virtual platform to enable the integration of "omic" data, analysis of algae genomes and associated bacteria metagenomics, the unveiling of metabolic pathways and phenotyping as well as the related necessary bioinformatics developments. Another aim is to promote seaweed production technologies to reduce the pressure induced by harvesting natural populations. Cultivating indigenous species, unmodified and non-pervasive, is one of the main criteria of the *Idealg* roadmap. The project encompasses an important effort to study the environmental impacts of harvesting and cultivating seaweeds, but also the impacts on society in terms of acceptability and economic activity of the coastal areas. What is at stake in this project is the integration of this high potential industry into a social, economic and sustainable development context.

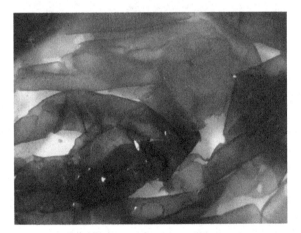

Figure 1.11. *The seaweed* Porphyra umbillicalis *is used in the preparation of sushi. It contains a sugar polymer cut by the prophyranase enzyme. This previously unknown enzymatic activity was detected in marine bacteria but also in bacteria that live in the intestines of Japanese people. The latter were probably exposed to the marine bacteria that contain porphyranase enzymes via their food. Genes would have been transferred from marine to intestine bacteria and this would have enabled the Japanese microbiota to acquire the faculty of degrading this sugar polymer,* © Murielle Jam/CNRS Photothèque 20100001_0626

1.2.9. *Genomic observatories*

Genomics observatories (GO) are first rate research facilities that produce genomic-level biodiversity observations that are contextualized, localized in territories and in compliance with international data acquisition standards. There are currently 15 of them. They represent marine and continental ecosystems for which genomic data acquisition is a long-term activity. These facilities aim to quantify the biotic interactions of ecosystems and to build models of biodiversity to predict the quality and distribution of ecosystem services. They are spread all around the globe: two in the Asia-Pacific area, including a French one in Polynesia - http://usr3278.univ-perp.fr/moorea/?lang=en, eight in Europe including the Rothamsted site used by the *TerraGenome* program and two French marine stations in Roscoff and Banyuls, involved in the aforementioned *Oceanomics* and *Idealg* programs - two in the Arctic and Antarctic Polar zones as well as three in the USA. They form a network that represents the "pulse of the planet" and whose main goal is to promote sustainable development through a better understanding of the interactions between humans and their environment. The approach consists of applying cutting-edge genomics technologies to monitor the stream of genetic variations in human and natural ecosystems. Genetic data is systematically related to biophysical and socio-economical data (metadata), which enables the integration of all the information into predictive models. Such models aim at mapping the quality and distribution of biodiversity as well as the ecosystem services it provides, according to various scenarios of future change and human activity. These observatories also play a major role in the promotion of training, technical support, resources and guidelines in the form of a learning portal. This internet resource is available for new sites or organizations that wish to perfom genomic observation and especially for new facilities from developing countries where the levels of biodiversity vulnerability are often high.

1.3. Acquisition, management and exploitation of samples and data

1.3.1. *Samples and collections*

Maintaining the integrity and availability of samples are critical issues for the projects we presented above, especially when the collections are gathered in a long-term perspective. The natural history museums are of course priority sites for this conservation task and nowadays they must not only keep the individual specimens but also their DNA. Beyond their patrimonial value, the collections and documentary resources they store are scientific resources of major importance to tackle the current issues about biodiversity. The collections of the French Muséum National d'Histoire Naturelle (The French MNHN is one of the largest collections in the world, two others being kept by the National Museum of Natural History in the United States (run by the Smithsonian Institution) and the Natural History Museum in London, at the basis of the *BarCode of Life*, a worldwide project that unites research projects that aim to implement barcodes* on the whole tree of life). Making collection specimens available to research projects requires expensive infrastructures both in terms of equipment and to make them valuable, for example through databases. France finances *e-ReColNat*, an online virtual national infrastructure to make these collections valuable via an image bank and collaborative tools for multimedia indexing, interactive content-based search and display functionalities. This project will become the largest virtual herbarium in the world.

Figure 1.12. *Sponges of the Homoscleromorpha class,*
Oscarella tuberculata, *off the coast of Porquerolles Island, France.*
They are used as a model for the study of chemical marine
ecology, © *Thierry Perez/CNRS Photothèque 20060001_1444*

The challenge in this context is to find the correspondence between genomic data and specific specimens. Where with naturalist collections, setting up such reference banks is generally possible, to associate a genome extracted from environmental DNA to an actual individual member of an identified species is impossible. This is because, as we saw earlier in this work, in samples collected in nature, whether they come from water, soil or air, one finds very large amounts of unknown organisms that must then be described. However, since the quantities are so great and the molecular biology era so developed – to the detriment of a morphological approach to organisms – we currently lack the specialists who would carry out this vast amount of systematic work, especially regarding the arthropods and organisms of small size, whether they belong to the terrestrial or planktonic domains. Taxonomic work must therefore also change scale and adapt its current ways to reach a "turbo" mode, in order to be able to process this avalanche of new biological data [THO 15, PAP 15].

Figure 1.13. *Samples kept before their mounting (herbarium sheets). Guyana Herbarium in Cayenne, © Claude Delhaye/ CNRS Photothèque CNRS 20090001_1625*

1.3.2. Production and analysis of data

Globally, environmental genomics projects face issues of interoperability and harmonization of environmental and genomic data acquisition and processing methods [CHA 15]. Data acquisition streams are becoming extremely fast and voluminous. The largest, most globally renowned sequencing centers (*NIH* and *JGI* in the US, *BGI* in China) are sized to achieve very large-scale sequencing campaigns for very large external projects that demand almost instant data availability. Beyond the molecular data that originates from "omic" approaches such as NGS, the data that these large projects have to analyze is also produced by measures and observations in ecology, imaging, geolocation, social networks etc. This massive influx of data in various formats [HOW 08] induces considerable methodological challenges in terms of data management, flow and sharing, as well as bio-computing analysis. Recent estimates predict that by 2025 environmental genomics and omics will have demands and needs that exceed those of the traditional big data providers that are astronomy, YouTube and Twitter [STE 15]. We urgently need to invest in technologies and pioneering research if we are to meet the environmental genomics challenges of the next decade. An inter- and multi-disciplinary approach is therefore necessary to promote and communicate at the highest levels the critical need for standardization of data and metadata* for their interoperability [TEN 14] with its corollary, data curation.

This avalanche of data raises the issue of its processing, storage and exploitation. Software development in bioinformatics and statistics must be reinforced and supported, as well as data storage computational facilities (services that can handle large amounts of data, parallel and grid computing, etc.). Open Data* [REI 11] raises a specific challenge because the large amounts of data it makes public often remain unused. What needs to be done is therefore to develop integrated models, not only of the distribution and diffusion of the measured data but also of the repartition of the observation effort. A new field of possibilities is also being opened by the recent developments in bioinformatics. It consists in analyzing the interactions (trophic network, inter- and intra-species competition, etc.) at various bio-geographical scales taking explanatory co-variables into account. The processing and exploitation of data require that the competencies of biologists and bio-computing specialists complement each other. This complementarity plays a major role in the success of research projects, as it fosters knowledge and know-how exchanges across disciplines. NGS data is typically fragmented, noisy and by definition not focused on a specific target (sequencing is a blind process). Its analysis requires the implementation of new models and algorithms that take into account genome redundancies as well as biases in the experimental designs that fast-rate technologies generate, while managing the constant changes of scale they induce. Currently, new methods proliferate, as each new issue is solved by an ad hoc local development, that is hardly transferable to other cases as it does not comply to any normative standard. An intense activity of methodological and software development in the perspective of standardization and interoperability must therefore be strongly encouraged.

The ultimate aspect of these new trends is the issue of interfacing and sharing data. This new working methodology is a true cultural revolution, whereby the main players of the research industry, as well as institutions, are encouraged to share data sets online (Open Data*) that can be exploited by others, and to make such individual contributions be beneficial in one's research career. A significant effort must be made to pool safe storage and access infrastructures in regional or national platforms in order to create open, normalized, curated and revised in the long run databases (see the *Ecological Data* and *DataOne* initiatives) and finally to apply minimal accuracy standards as defined by international consortia (*Ecoinformatics*).

1.4. Tomorrow's environmental genomics

Biology is now addressing considerable issues that were unimaginable only 10 years ago. Technological advances have occurred in genomics, of course, but also in imaging, hardware, monitoring and experimentation at all biological and spatial scales. This enables the construction and analysis of immense data sets, and to relate genomes, transcriptomes and proteomes with corresponding phenotypes and their associated metadata. There is a strong demand from society to curb environmental impacts and anticipate consequences in the short, middle and long run. Therefore, research must consider the complexity of biological systems, including humans, to be able to elaborate scenarios that predict their dynamics in time and space. Approaches such as data mining, which consists of extracting information from vast data sets, must be intensified in all disciplines, from biodiversity to clinical genomics. A specific emphasis must be put on the synthesis of observations, and relating them to the other levels of organization. Interdisciplinary and multidisciplinary strategies are therefore major assets to magnify the various types of expertise, especially in modeling. This is necessary in order to meet the challenges created by the paradigmatic shift in biology, whereby the main difficulties arise from the identification and selection of the sheer number of parameters that must be considered.

2

Technological Revolutions: Possibilities and Limitations

The 21st century has seen a revolution in DNA sequencing technologies. After decades devoted to improving the first-generation DNA sequencing methods, major technological revolution has enabled the emergence of second- and third-generation DNA sequencing methods, and they have quickly become prominent. These new technologies are still in their early stages when advances are expected as the quantity and accuracy of the produced DNA sequences still need improvement.

2.1. The technological revolution of second- and third-generation sequencers

First-generation sequencing using the Sanger method [SAN 77] is based on the insertion of modified nucleic acids; this step is followed by the physical separation of DNA fragments of various sizes and then by their optical detection. This three-step process enables the reconstitution of the sequence of DNA nucleic bases.

The emergence of second-generation sequencers is due to a major technological revolution [GRA 07, LOM 12]. This improvement enables the sequencers to amplify DNA-fragments on a fixed support and to read the DNA-sequence by optical detection, simultaneously. Hence, this dual functionality allows the system to gain several orders of magnitude in the quantity of reads that can be produced. While 384 reads of an average of 900 bp may be produced in 16 h on a first-generation sequencer (ABI 3730 XL sequencer) that was used in 2004, more than 10^9 reads of 100 bp are produced in 12 days on a second-generation machine (Illumina HiSeq2000 sequencer). Various types of second-generation sequencers are proposed on the market, with various characteristics regarding the accuracy, quantity and read lengths of DNA sequences, as well as with various modes of operation (optical for Illumina and 454, electro-chemical for Ion Torrent) and DNA sequence reading speeds. NGS are still emerging technologies and therefore the technologies and hardware they are based on are still rapidly evolving. Roche Diagnostics, who bought 454 Life Sciences in 2007, announced that services related to 454 sequencers would be put to an end during 2016. Nowadays, Illumina sequencers are the most widespread worldwide [FAU 15]. Other technologies are currently being developed or are even already on the market: Heliscope single molecule sequencing by Helicos Biosciences, DNA nanoball sequencing by Complete Genomics, Nanopore by Oxford Nanopore Technologies.

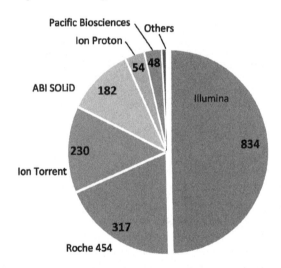

Figure 2.1. *Distribution of new sequencing technologies in sequencing centers across the world (December, 2015). The figure shows the number of sequencing centers that own each individual NGS technology. Data is based on 7,389 machines distributed in 1,027 sequencing centers. Data available on the website www.omicsmaps.com/stats*

A new technological breakthrough has recently been made: the DNA amplification step is no longer necessary. Nowadays, third-generation sequencers can sequence single DNA molecules that are more than 1000 bp (= 1 kbp) long (up to dozens of kbp). Since 2011, Pacific Biosciences has been commercializing DNA sequencers that use single molecule real-time sequencing (SMRT) technology: a DNA polymerase is attached at the bottom of a sink (~70 nm in diameter) that only one DNA molecule can access; the insertion of each nucleotide (A, T, G or C) is measured by fluorescence emissions. Innovations in the field of reactives improved the average size (7 to 10 kbp) of the obtained sequences. Other innovations managed to increase the number of sinks per SMRT cell: from 150,000 sinks for the PacBio RS II sequencer up to 1 million for the following system sequencer, available since 2015. These innovations improved the accuracy and quantity of obtained DNA sequences. This technology can be used for *de novo* genome sequencing but also for epigenome* analysis, which refers to the detection and methylated genomes, because the activity of the DNA polymerase is sensitive to DNA modifications.

2.2. Genomic DNA sequencing from a single cell: "single cell genomics"

Genomic DNA sequencing from a single cell, "single cell genomics" (SCG), is an innovative use of NGS.

SCG is achieved through several steps:

1) cell isolation by dilution, micromanipulation or automated sorting by flow cytometry;

2) lysis of cells;

3) WGA (Whole Genome Amplification)-based amplification of DNA, mostly using the DNA polymerases Phi29 or Bst with the "multiple displacement amplification" (MDA) method;

4) NGS sequencing. This original method still suffers from some defects and biases. One of them is linked to difficulties in the lysis, i.e. in the disintegration of the cell wall of organisms. Another is the biases induced by the MDA amplification, as it generates DNA chimeras. Some of these biases can be reduced by decontaminating the reactives, miniaturizing and, therefore, reducing reaction volumes from nanoliters (10^{-9} liters) to picoliters (10^{-12} liters).

Completion rates of the resulting genomes are variable (0 to 100%) and depend on many factors such as the amount of available DNA after the cell anlysis, the contamination by other DNA sequences and the genome complexity. The SCG approach is expected to enable the identification of rare or still unknown organisms from one single cell. This would give access to non-cultivable organisms, to the characterization of infra-specific biodiversity to document genomic variability or the inter-specific variability within communities of organisms and even to the study of their evolution and adaptation by detecting mutations within populations of organisms. Other fields of research that the SCG approach opens up include studies on epigenomes, embryology, organogenesis and neurobiology as well as medical and clinical studies on humans, related for example to research on cancer processes [WAN 15].

2.3. Availability of NGS technologies to laboratories

Public or private research laboratories usually have their own second- or third-generation sequencers if they do not use regional or international public or private sequencing platforms. The current market is characterized by a great diversity and by intense competition.

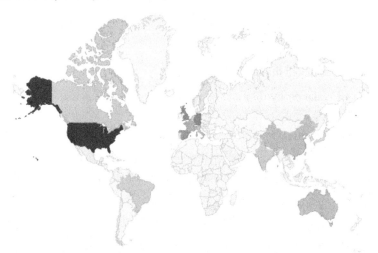

Figure 2.2. *World map of the distribution of sequencing centers in countries (data from http://omicsmaps.com/stats, December 2015). Color intensity grows with the number of centers (1 to 302 in the United States). For a color version of this figure, see www.iste.co.uk/faurejoly/genomics.zip*

Large sequencing centers develop very large scale sequencing programs. In France, France Génomique, for example is developing projects that aim to unveil the vegetal biodiverstiy of the Alps or that of oceans' eukaryotes or of soil microorganisms; the Beijing Genomics Institute (BGI) is involved in the sequencing of the genomes of 1,000 humans, 1,000 animals and various plants of scientific and economic interest while the Joint Genome Institute (JGI) proposes a sequencing project of 1,000 fungal genomes, or a project towards metagenomics of microbial communities of hot springs and soils (see Chapter 1).

2.4. NGS data storage and processing

Computer processing and storage of large flows of NGS data lead to significant constraints. A HiSeq2000-type sequencer can for example produce almost 1 TB of data in two weeks. Current estimates give a ratio of 10 to 100 between the costs of analysis and storage and the costs of actually producing the sequences. Analysis and storage are thus becoming more expensive than producing the data in the first place. Nevertheless, keeping reference metagenomes, genomes or genetic sequences in databases is essential, as they constitute the archives of the knowledge about environmental genomics. Conversely, in some cases, prioritizing the conservation of environment samples over computational NGS data can be more appropriate because DNA sequencing data of a better quality may be expected in the near future.

3

NGS Data Sharing and Access

Environmental genomics provides the scientific community with ever larger data sets concerning an ever increasing number of organisms and ecosystems. Thanks to a parallel increase in computing power and software efficiency, these new data sets enable in-depth analysis to a level that was not possible ten years ago. However, this avalanche of data, which is at the basis of the NGS success, is so dramatic that it raises challenging issues about its sharing and access.

3.1. The large DNA data banks

For the past 30 years, the three large generalist data banks that collect (meta)genomic and (meta)transcriptomic sequences have been *GenBank* at the *National Center for Biotechnology Information* (*NCBI*), the *European Nucleotide Archive* (*ENA*) at the *European Bioinformatics Institute* (*EBI*) and the *DNA Data Bank of Japan* (*DDBJ*) at the *National Institute of Genetics* (*NIG*). Although their sizes and contents were initially quite different, an international cooperation strategy was quickly set up. For the past 25 years, their content has been virtually the same: any sequence submitted to one of them is forwarded to the other two within a day. The main advantage of these three banks is that they enable free access to almost all the biological sequences that have been produced by public as well as private laboratories.

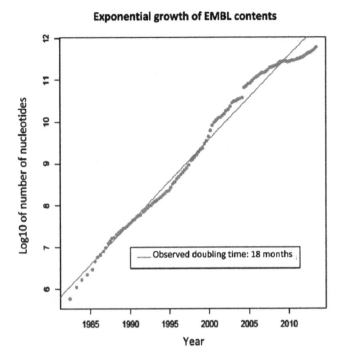

Figure 3.1. *Exponential growth of EMBL
contents, © Guy Perrière*

3.2. Constraints on access to DNA data banks

Before the advent of NGS technologies, the institutions in charge of the data banks (GenBank, ENA, DDBJ) could manage the flow of data even if it was steadily increasing. This is not the case anymore, because the amount of data communicated to the three collections has grown exponentially since the start of the 21st century, its amount doubling every 18 months. Since 2010, however, this trend has changed, as a lengthening of doubling time has been observed. This unexpected phenomenon can be explained by the fact that centers responsible for the maintenance of these banks have had increasing difficulties assuming the financial and technical challenges involved with the processing of so much data as the continuous purchase of storage capacity and the maintenance of associated structures have become major challenges.

Another issue stems from the fact that the amount of produced data is such that transferring it to data bank centers via the internet has become impossible. A common practice nowadays is to transfer them on a physical support (like a hard-drive) via the post. It is a paradoxical come-back of pre-internet practices as, until the end of the eighties, sequences were communicated by physical supports such as magnetic bands or floppy disks sent through the post.

Finally, the access to short, non-annotated reads* also raises significant challenges. As so many sequences are available, the centers do not propose access to individual sequences anymore, they only provide access to compressed archives that gather together many reads. Cost concerns have often endangered the survival of these archives, as when, for example, the service was interrupted in EBI. It was later re-established.

In this context, more and more sequences are simply never sent to the data banks. Their availability to the community is mediated by local collections set up by limited projects. As a consequence, there exists nowadays, in addition to the three large generalist collections (GenBank, ENA and DDBJ), a plethora of smaller banks that specialize in one organism or in one specific biology issue. The journal *Nucleic Acids Research* publishes a special issue each year dedicated to the main banks available in the world. In its 2016 edition, no fewer than 1685 banks were listed (volume 44, issue D1, January 2016). This inventory is far from exhaustive and the number of specialized banks is probably much higher.

3.3. Computing architectures for NGS data

Two levels of computing infrastructure must be considered for genomics (e-infrastructure): the infrastructure that deals with raw data, and the infrastructure that deals with computing processing on the data, producing filtered data with a higher added value because it can be exploited by scientific communities. These two levels require similar storage volumes but the latter needs much more significant computing and temporary data creation resources.

Sequence data production is currently decentralized to several local sites distributed across several countries, but also across some private service platforms. The fact that the technologies have become easily available to smaller groups means that the amount of data produced has soared, but that the number of production sites has as well, which entails the risk of seeing computing means and know-how fragmented and duplicated. More coordination in terms of infrastructure and investment seems necessary.

Figure 3.2. *Computing center, © Cyril Fresillon/CNRS Photothèque 20120001_1607*

Computing needs for the processing of massive sequencing data include fast computing high performance computing (HPC) as much as high throughput computing (HTC). Surprisingly, the national computing centers are still rarely used by the genomics community, although grid computing was an option explored very early on by the French bioinformatics community. The GRISBI initiative, "Grids for bioinformatics", supported by the RENABI network, GIS IBISA and GIS France Grilles, enabled research to assess how grid computing could help to address bioinformatics issues more harmoniously and collaboratively, in case, for example, of the management of data produced by new-generation sequencing projects. The study highlighted the necessity to address several issues. The first is the issue of storage and data archives. We must be able to transfer, duplicate and modify data on various sites: production site, archiving site, analysis site, site for exploitation and visualization by biologists. Technologies such as *iRODS* enable virtual storage on several sites, thus providing simple remote access and sharing of data thanks to a relevant metadata management policy. The second issue is related to the quality of the network, as we must ensure that data is communicated in a fast and secure manner to all sites involved. The third and final issue concerning the exploitation of high throughput sequencing data is the necessity to provide biologists with the ability to run processing chains on grids, supercomputers or clouds, according to their needs. Collaborative platforms such as *Galaxy* or *R* are making this possible.

3.4. Standards in genomics

Chapter 1 informed us that the amount of data produced doubles every 6 months, thus following Moore's law about the relationship between data amounts and necessary data storage space. The era of "BIG DATA" [DAT 08] leads to various analytical strategies that go beyond the capacities and interests of individual researchers [TEN 14] or government or non-government institutions. Science is thus entering a new paradigm in which it is not driven by scientific questions anymore, but by the way data is organized and analyzed. A critical issue is then that of developing methods that are compatible with the sharing of data as well as with its associated metadata. During the past decade, several attempts at developing methods for the conservation and management of genomic data [FIE 09], the design of policies for sharing, for information requirements, for terminologies, as well as for models and exchange formats have been tested. The purpose behind these attempts at normalizing high throughput data is to maximize its interoperability. This international effort led to a proliferation of norms, guidelines and formats that may look like an impenetrable jungle to the eye of the young researcher, a jungle in which even the most experienced can get lost. In this context, the policy of "if it isn't broken, don't fix it" often prevails.

Several players play a significant role in the spreading of the norms of data sharing: first of all, the publishers and financing bodies can impose norms upon submission or funding of works or programs about genomics [KAY 09]. Some of them (Nature Biotechnology, BioMed Central) actually stirred up a debate whereby the community tried to identify the best practices [SAN 12]. However, like most new developments, these initiatives had a limited impact on the daily scientific practice because the instructions for authors are not always grounded on solid bases or on equivalent norms between editors and agencies. The requirements of scientific journals vary greatly, ranging for example from sharing the primary data of the research, to sharing to only those that request it, as it is stated in the paper that the data can be provided on demand [ALS 11]. Out of 500 reference articles in the field, 30% do not have to comply to a data sharing policy and 59% do not comply with data availability requirements. This shows that there remains ample room for flexibility in the implementation of sharing and availability. This leeway should be overcome to ensure data sharing whilst protecting intellectual property rights.

Figure 3.3. *The Curie supercomputer, owned by GENCI and operated into the TGCC by CEA, is the first French Tier0 system open to scientists through the French participation into the PRACE research infrastructure. Curie is offering 3 different fractions of x86-64 computing resources for addressing a wide range of scientific challenges and offering an aggregate peak performance of 2 PetaFlops, © Cyril Fresillon/CNRS Photothèque 20120001_0230*

3.5. Metadata

Metadata is to the digital world what a note is to paper-based documents; it gathers descriptive (title, author, date, subject, editor, etc.,), semantic (keywords, summary, descriptors, etc.), structural (organization of text and data) and administrative (date and context of production, formatting type, access conditions, rights and intellectual property, archive and other technical details) information. Metadata describes data at all levels of aggregation (collection, single resource, part of a larger resource, etc.). It can either be integrated in the data (often in XML format) or accessible separately.

With metadata, the notion of interoperability of data, i.e. the possibility for computing systems (hardware and software) to access, exchange and query the data becomes a major issue. Within the framework of the semantic web, which favors common data exchange methods, metadata enable the use of data in contexts different from what they were originally produced for. The existence of metadata participates to the emergence of new forms of knowledge by relating and structuring information on the data itself.

Many private solutions, imagined by individuals or the private sector are being designed to optimize and automatize data and metadata querying methods, according to the specific contexts and applicative domains in which they are created. Interoperability then raises the issue of standards and formats as well as that of controlled lexicons (who, what, when, where and how). The validation of such standards at the international level is defined by ISO norms, but they are specific to each application type (bibliographic, geographic, technical, molecular, etc.).

In the context of genomics and even more so in that of environmental genomics, metadata must provide clear and easily accessible information on content, structure, context and conditions in which the data was collected and analyzed. As detailed by Bild *et al.* [BIL 14] in their field guide for research in genomics, genomic data sharing practices must include:

– a clear and detailed method description, so that others can apply them independently from the main goals of the owner;

– availability of raw and processed data in public stores, like for example in Gene Expression *Omnibus* or in the Database of genotypes and phenotypes, or as sequence of reads saved in archives;

– sharing of code and scripts in version control reference platforms such as *GitHub* or *SourceForge*;

– and, lastly, access to code files or scripts that analyze the data.

In 2005, the *Genomic Standards Consortium* (GSC) [FIE 11] was created. It is devoted to defining norms in the field of genomics. It is run by a community of genomic scientists whose aims are to improve the quality of the context descriptions of sequence data. The consortium favors the normalizing of genome descriptions, as well as exchanges and integration of genomic data. It specifically focuses efforts on biodiversity genomics in the framework of the *Genomic Observatories* (*GO*) network, which is devoted to the long-term monitoring of the environment on various sites across the world. GSC is an open international community that gathers about 200 researchers that cover several fields such as biology, bio-computing and computer science. It also includes representatives of international nucleic data bases such as *DDBJ*, *ENA* or *GenBank*, as well as representatives of major sequencing centers such as the *Argonne National Laboratoire* (*ANL*), the *J. Craig Venter Institute* (*JCVI*), the *Joint Genome Institute* (*JGI*), the *Institute for Genomic Science* (*IGS*) and the *Welcome Trust Sanger Institute* (*WTSI*).

The GSC created the "minimum of information about a sequence" (*MIxS*) norm that gathers three types of lists of minimal information about genomes (MIGS), metagnomes (MIMS) and sequences of sequence markers (MIMARKS). These lists detail environmental metadata variables in normalized formats for the myriad environmental variables that correspond to the various specific characteristics of the situation (ammonia concentration, conductivity, wind speed, patient health, etc.). The GSC also works towards the development and maintenance of adaptable tools and interfacing services and towards reliable archive solutions with a focus on taxonomy and biodiversity. This is why the consortium remains open to new members to support the design of shared responses that integrate the diversity of the environmental contexts of various research programs.

The compliance with these standards and the recording of researchers' data and metadata in public bases is supported and even increasingly frequently required by scientific journals.

3.6. Conclusions

As an increasing amount of sequences are no longer sent to the three generalist collections, GenBank, ENA and DDBJ, these banks cannot be seen as exhaustive anymore. The end of that exhaustiveness is already influencing the way bio-computing analysis of DNA data is done. With the aim of addressing this issue, *EBI* launched, at the beginning of 2007, the *ELIXIR* initiative. This initiative aims to federate the most important bio-computing centers (national and regional centers) in a European network.

4

Accuracy of NGS Data: From Sequence to Databases

Second-generation sequencing led to the production of DNA sequences with considerable throughput rates, but with reduced read lengths and varying error rates. This avalanche of sequences requires the development of new bio-computing tools to ensure that the information is optimally processed [LOG 12]. Requirements for archive storage space and computing power are multiplied, which leads to the deployment of new computing infrastructures. The quality and accuracy of the data produced and the relevance of their analysis are now the challenges at stake for environmental genomics.

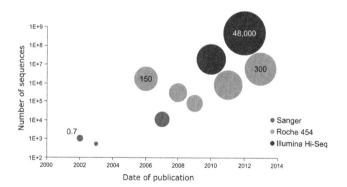

Figure 4.1. *Unprecedented inflow of NGS data: example of the viral metagenomes published in the literature. For each study, the number of sequences (y-axis) and the quantity of base pairs (in millions) are depicted as a function of the publication year (x-axis), © Denis Faure. For a color version of this figure, see www.iste.co.uk/faurejoly/genomics.zip*

4.1. Accuracy of NGS data

NGS sequencing is done through a process that requires manipulating the DNA extracted from the organisms, for example by constructing and amplifying the libraries and by the addition of adapters at the ends of DNA fragments. The added sequences are used to amplify, sequence and recognize the samples studied. Bio-computing algorithms must then not only eliminate these additional sequences but also detect the segments of sequences of poor quality, identify isolated sequencing errors and artifact sequences that are due to the biological manipulation. Not taking these inherent defects of the raw sequences produced by NGS technologies into account would undermine subsequent processing steps such as sequence assembly* which are necessary to produce complete genes and genomes, their annotation and biodiversity appreciation.

4.2. Accuracy of taxonomic affiliations

The accuracy of taxonomic affiliations is a methodological and scientific bottleneck when one intends to understand the diversity of organisms. Molecular identification of organisms involves analyzing phylogenetically informative genes (for example the *rss* gene expressing the 16S or 18S rRNAs in prokaryotes and eukaryotes respectively) that can easily be isolated and that enable the determination of evolutionary relationships. This approach, which consists of identifying organisms with a genetic marker, is often called barcoding (see Chapter 4). Taxonomic assignment on the basis of a sequence can be achieved by merely looking for sequence similarities (a quick but approximate method) or through the construction of phylogenic trees on the basis of evolutionary models and of a significant number of reference sequences (a computationally longer but more accurate method). Since the computational costs associated with the construction of phylogenic trees on the basis of NGS data are great, the similarity search approach remains favored, which sometimes leads to erroneous assignations or the lack of assignation (due to too weak similarities with reference sequences). Recently developed methods that enable researchers to place query sequences on a reference phylogenetic tree should make this approach easier to conduct. Furthermore, the use of several phylogenic markers or even of complete genomes should pave the way to a more resolutive phylogenomics that would be able to detail identifications and lineages of organisms. This accurate approach was recently applied to environmental data [CHI 13].

4.3. Accuracy of genome and metagenome annotations

The accuracy of genome annotations (gene identification and prediction of the proteins these genes encode) suffers from the exponential growth of the number of sequenced genomes. Our present knowledge remains paradoxically insignificant in comparison to the amount of data that is produced and to the diversity of the living world. Genomic data are typically stored in public repositories but remain under-exploited, even though the science of data-mining is undergoing new advances. One of the factors that still restrains the exploitation of genomic data is the difficulty of predicting gene structure, which remains one of the most important unsolved issues of computational biology. Such predictions can be achieved either by comparing sequences with known genes of various organisms or by using *ab initio* predictive algorithms that are based on the intrinsic characteristics of analyzed sequences. These methods, despite recent significant advances, are still not able to generate an exhaustive inventory of all the genes. To improve our predictive ability, we should consider devising more precise rules to determine whether segments of DNA contain genes or not.

The analysis is even more complex in the metagenomic studies of microbial communities. Indeed, in that case the task does not consist of analyzing an organism or a homogeneous population of organisms but of simultaneously analyzing sets of dozens, hundreds or even thousands of species. The extraordinary diversity of the microbial communities of most ecosystems, therefore, makes it difficult, or even impossible, to reconstruct complete genomes on the basis of short-length sequences, even if a massive amount of them is produced. Despite constant progress and new algorithms, assembly software and computing power are currently not efficient enough to achieve the *de novo* reconstruction of all the genomes of a community [AFI 12]. Furthermore, since millions of reads remain non-assembled, identifying genes within these small sequences remains problematic. Despite these limitations, there are some very popular metagenomic data processing software solutions (such as MEGAN) that can provide automatic sequence annotations and that can also, thanks to statistical analysis methods of various complexities, compare (meta)genomes. Nevertheless, since annotation accuracy is crucial to the constitution of reference data, significant efforts must be invested in the development of efficient and accurate genome and metagenome annotating strategies. Accurate reference data should thus be generated and classification approaches that make valuable the sequences of still poorly characterized organisms.

Figure 4.2. *Data accuracy. Research towards the understanding of biodiversity uses NGS technologies that provide a data deluge that must be processed by innovative bio-computing strategies for it not to remain underexploited, © Pierre Peyret and Martine Chomard. For a color version of this figure, see www.iste.co.uk/faurejoly/genomics.zip*

4.4. Quality of the databases

The quality of databases is essential to make NGS data valuable. The main resource of nucleic data that enable the sharing and widespread distribution of knowledge (the International Nucleotide Sequence Database Collaboration) originates from a collaboration between Japanese (DNA Data Bank of Japan), American (GenBank) and European (European Nucleotide Archive, see Chapter 3) databases. Such data collection strategies require the definition of standards with which submitted data must comply for sharing and re-use to be efficient. Such standardization initiatives have been led by the Genomic Standards Consortium (GSC) for genomic and metagenomic data and by the Consortium for the Barcode of Life (see Chapters 1 and 6) that aims to set up a standardized library of all living organisms. Practically, the Barcode of Life (BOLD) provides a web platform enabling the recognition of an organism via comparisons with its database. The project, funded until 2015 led to a bank of 5 million specimens covering half a million species, which are for the most part linked with their geolocalization and with the Natural History Museums that store them. Specialized databases were also developed to identify prokaryote microorganisms (the Ribosomal database project-RDP, Greengenes, SILVA). So, many specialized databases flourish, whether they focus on a phylogenic, (meta)genomic or metabolic analyses, but the rising amount of sequence data requires new computing advances to ease transfers, security, storing and analysis.

Figure 4.3. *Water meadow. Only high quality databases will enable relevant taxonomic information to be related to NGS data that detail the biodiversity of ecosystems, © Gregory Loucougaray/CNRS Photothèque 20120001_1071*

4.5. Some best practice guidelines for the analysis of NGS data

To produce relevant results on the basis of the masses of data produced by NGS, specific computing and statistical tools are of course necessary, but a good preliminary reflection is also required. Here we discuss the identification of genes that are differentially expressed depending on a variety of contexts, on the basis of transcriptomic data obtained by RNA-Seq (number of copies of cDNA that correspond to RNAs). The same approach can be applied to the identification of taxons whose amounts differ according to the context, on the basis of barcoding counts by marker gene amplicon sequencing (the *rss* gene, for example), provided that most taxons are present in all the compared contexts.

From data generation to its processing, each stage has significant consequences for the following stages. Data normalization aims to detect technology-induced biases and to amend them for the resulting samples to be comparable with each other. It is therefore specific to each technology and platform and is a very sensitive task because it involves modifying raw data. It must, therefore, remain strictly within the bounds of what is necessary. Some biases can be eliminated by an appropriate experimental strategy, a well-chosen experimental protocol and efficient bio-computing processing. The main bias is the differences in sequencing depths (total number of reads) of samples. In that cases, methods based on effective library sizes, defined from regions of interest that vary little from one sample to another, seem the most appropriate [DIL 13]. They remain efficient even if the RNA repertories are expressed in very different ways. If biases of the "specific sample" type due to the concentration in GC bases of samples are observed, an additional normalization can be necessary.

Differential analysis consists of detecting the statistically significant zones, according to a chosen threshold, with the help of statistical testing. The threshold depends on the number of tests. RNA-Seq specific methods, including DESeq2 [AND 13] and edgeR, implemented in the packages BioConductor of R, were developed in the perspective of a low number of replicates (less than 5) in each context.

Searching genes that are expressed differently in two or several transcriptomes remains a challenge because discerning whether variations denote various species or the intra-species variations at the level of gene expression is still difficult nowadays.

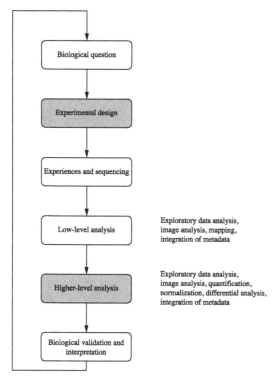

Figure 4.4. *Guidelines for NGS data analysis, © Julie Aubert*

4.6. Conclusion

For the amounts of data produced not to be detrimental to its quality, increasing the efforts in interdisciplinary research that associates biology, computer science, bio-computing and mathematics is the obvious strategy. For that knowledge acquisition revolution to keep developing, expertise is a key factor for the success of quality data production, administration and analysis and this entails a training appropriate to the current issues at stake. Technological advances (third-generation sequencing, cell isolation, gene capture, computer infrastructures, etc.) together with the development of new algorithms will certainly keep progressing and accelerating. This should enable the reconstruction and analysis of numerous reference (meta)genomes and (meta)transcriptomes which, in turn, will propel a new perspective on the living world, a perspective that takes interactions and exchanges between organisms into account.

5

Taxonomy and Biodiversity

Biodiversity is a concept to describe the structure and dynamics of life not only at species level but also at infra- and supra-species levels. The two following examples highlight the importance of the concept of biodiversity: there is 5 to 10 times more genetic divergence between two chimpanzees of the *Pan troglodytes* species than there is between two of the most distant humans; generic variations within a single species of angiosperm are on average much greater than genetic variations within the species of coleoptera. Furthermore, species interactions and associations between species that co-exist locally create communities that structure themselves at their own scale. In short, counting species on the basis of taxonomic criteria is the first stage of an exhaustive account of biodiversity. It must be furthered by the analysis of phylogenetic relationships and interactions between organisms and the environment.

5.1. How can we measure biodiversity?

Biodiversity can be understood in terms of function assumed within ecosystems to the point when one can assess the ecosystemic services it contributes to. This functional approach of biodiversity enables its integration in an economic, legal and social paradigm. This is a legitimate perspective, as society's first demand is to predict how anthropogenic changes, since they disrupt the dynamics of life, will alter the services that ecosystems provide for humans. However, beyond the services and functions are the species. To prioritize the ecological role of biodiversity entails the risk of reducing biodiversity to limited numbers of functional groups, whereas recent research on ecosystem functioning demonstrates the importance of rare species, which have remarkable specialization but which are functionally vulnerable.

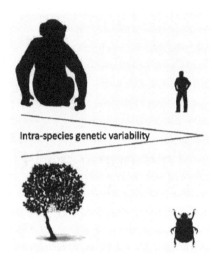

Figure 5.1. *Genetic variability is more significant in chimpanzees than in humans; it is more significant in angiosperms than in Coleoptera beetles, © Sylvie Salamitou*

Another way to understand biodiversity is through its historical dimension, whereby the originality of the structures and figures of organisms is described in terms of inheritance. In the African bush, for example, the pangolin and aardvark are both predators of social insects, thus they provide the same ecological service. However, in terms of evolutionary divergence, pangolins are much more interesting than aardvaks because the latter have very close cousins in the same environment (damans, elephants, elephant shrews). The ecological order is not systematically the mirror of the historical order.

Classifications are precious tools for systematicians, but also for many users such as system ecologists, managers of protected areas and for many actors in the industry in fields as varied as cosmetics, pharmacy and agribusiness. It is now possible to relate the presence, lack or abundance of actual precisely identified species and the communities they constitute to environmental parameters, thanks to the development of models of diversification and niches. It is also possible to apply such studies to fossils and paleo-environments to learn valuable information for the prediction of future biodiversity dynamics when faced with the return of similar environmental conditions [CON 13].

Figure 5.2. *Workshop for the inventory of marine diversity during the KARHUBENTHOS expedition in Guadeloupe (France). The challenge here consists of describing biodiversity whilst ensuring that the links between each specimen of the collection and the DNA sequences of his genome are established,* © *Line Le Gall*

5.2. Taxonomy in the NGS era

Describing the whole of the biodiversity on the planet for a long time seemed almost unrealizable, especially because a massive, rapid extinction called the sixth extinction has already begun. However, a recent estimate of biodiversity at 5 to 8 million eukaryote species [COS 13] means that its description is actually possible using NGS in association with novel practices in taxonomy studies. Furthermore, thanks to NGS, new aspects of biodiversity have become accessible. For example, the genomes of microorganisms that cannot be cultivated in laboratory conditions can now be identified and assembled with metagenomic approaches [IVE 12]. The essential issue is, therefore, how to enable communication between the fields of biological systematics and environmental genomics. The two approaches could mutually enrich each other and, thereby, maximize the knowledge we have about every taxonomic entity.

NGS technologies are particularly promising for the analysis of the composition of communities, especially the communities that are threatened by rapid extinction due to global change. The risk is, however, that they may build a taxonomy of OTU* that is independent of the taxon names and associated knowledge. This OTU analysis approach may seem acceptable for some quantitative evaluations of biodiversity, especially when we are faced with emergencies (deforestation or changes in land use, for example). In other cases, if we want to understand the origins and dynamics of the biodiversity of a biotope or a specific ecosystem, we must favor a taxonomic metabarcoding approach. That means ensuring that NGS sequences are related to taxonomic names and associated knowledge thanks to "voucher specimen" collections.

Figure 5.3. *New molecular taxonomy NGS approaches enable researchers to describe the variations of communities of organisms under anthropic pressure with unequalled taxonomic accuracy. Here, slash and burn deforestation, "Baobab" region in Kirindy, Madagascar, © René Bally/CNRS Photothèque, 20100001_0922*

The merging of NGS and taxonomy practices combines various fields of expertise in compared anatomy, taxonomy, bio-computing (as there is a considerable volume of data to process, store and share, including data from new imaging technologies), logistics of *in situ* biodiversity exploration and the conservation of specimens in Natural History collections, which are becoming major research infrastructures.

5.3. Methodologies for taxonomic identification using NGS

The phrase "taxonomic identification" covers two central approaches to taxonomy that are very different as they call upon different methodological categories:

– on the one hand approaches that assign unknown specimens to already defined taxonomic groups (for example a species): this is called "specimen identification";

– on the other hand, approaches that delimit species or that more generally perform phylogenic inferences for levels higher than species: this is called "species discovery and species delineation".

Before turning to the methodological aspects of these two approaches, we must remind the reader of the crucial importance of taxonomic sampling because these two approaches rely on a relevant evaluation of intra- and inter-species variability (or more generally, intra- and inter-group variability). Importantly, all the currently available variability evaluation methods are sensitive to the quality of sampling. In order not to underestimate intra-group diversity nor to overestimate inter-group diversity, a sufficient coverage must be ensured not only within each group but also between groups as a whole. For a specific level, this consists of:

1) covering the whole of the area on which the species is spread and including in the data set several specimens for each population;

2) ensuring that all the species of the studied genus are included in the data set. The direct consequence of these sampling requirements is the necessity to design NGS sequencing strategies that enable the taking into account of numerous specimens to analyze.

The community of molecular taxonomists is very active in the definition of new methodologies. Beyond the now classical phylogenetic approaches, ("total evidence", "supertree"), they regularly propose new methods of species assignation or delimitation. These methods may use a single locus ("monolocus" methods) and are in that case fit for data sets with many specimens; they otherwise may use several to thousands of loci (multilocus) and are in that case often limited to samplings that include only a limited number of specimens. Thanks to NGS, it is now possible to combine the benefits of both types of methods (many markers and many specimens) as the approaches like RAD-seq do for large eukaryote genomes or for the full sequencing of genomes of bacteria, archaea and viruses.

One constraint is that some NGS methods often lead a greater error rate than the first generation sequencing methods of Sanger. Even if the NGS techniques keep progressing and reducing this error rate, such potential biases must be taken into account in the experimental design by carefully balancing the number of sequenced loci, the number of sequenced specimen and the minimum coverage of each sequenced fragment.

5.4. Microorganisms: the present challenges to modern systematicians

The common term "microorganism" covers a set of very different taxonomic groups: some eukaryotes like fungi, but also bacteria, archaea and viruses.

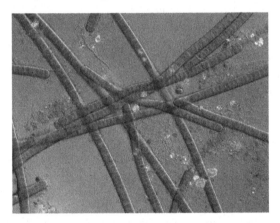

Figure 5.4. *Overview of a complex microbial community in an aquatic environment (Orsay, France), © Purificacion Lopez Garcia/CNRS Photothèque, 20080001_0109*

Two distinct processes are jointly responsible for the great genetic and genomic diversity of these organisms: a process of heredity (vertical transfer) whereby the parents' genetic heritage is transmitted with a few variations; a process of acquisition of genes from other species (horizontal transfer) through various mechanisms (introgression in dimorphic organisms, but also conjugation, transduction and transformation for bacteria or archaea). The phenotypes of microorganisms are, therefore, partly independent of vertical heredity: phylogenetically close organisms can exhibit different traits when they have acquired different genes by horizontal transfer; conversely, organisms that are phylogenetically distant can exhibit similar traits if they received similar genetic material via horizontal transfer. These transfer processes mean that the genomes of many lineages of microorganisms are phylogenetically composite: not all the genes of a genome come from a single common ancestor they come from various sources in varying proportions.

One remarkable consequence of this genomic fluidity is that the members of the same bacterial species do not all have the same families of genes. The gene families which are common to all the members of that species constitute the "core-genome". For a given species, the total sum of all gene families, which is larger than the number of gene families of any single individual is named the "pan-genome". Modeling the phylogeny of such composite taxonomic entities by a phylogenic tree is, therefore, ill-suited because the tree formalism suggests a single ancestor for organisms whose individual parts have various histories, each linked to at least one ancestor.

A reductionist approach consisting of sequencing a single locus to tell the evolutionary history of a species nowadays seems very simplistic to describe the origins of microorganism biodiversity. Not all genes have the same history and the history of one single gene is not the same as that of the species. Genomic NGS and "single cell" methods, applied to environmental communities, can now be used to better describe the gene collections in microbial communities to infer taxons, horizontal transfers, interactions and adaptations in the microbial world.

Figure 5.5. *Extraordinary marine biodiversity: gorgonian, red coral, sponges, crustose red algae forming a coralligenous reef structure in the Mediterranean Sea, inhabited by barbiers* (Anthias anthias) *and other fishes, © Roland Graille/CNRS Photothèque, 20070001_1114*

5.5. Towards integrative taxonomy

Integrative taxonomy nowadays creates new opportunities for dialogue between pre-NGS taxonomy approaches and approaches that integrate NGS, such as the multilocus barcode [CRI 14]. Important research to test new hypotheses on the delineation of species is becoming possible. Integrative taxonomy not only takes DNA sequences or barcodes of the specimen type into account, but also various characteristics that can be anatomical, morphometrical, biochemical, ecological, behavioral, etc. (for examples, see the ITIS website). Furthermore, inclusion of old specimen collections become possible because the application of NGS to ancient or degraded DNA enables the production of sequences on the basis of an extremely low volume of holotype material [PUI 12].

Although the integrative approach is now widely accepted by taxonomists to delimitate taxons, its application often remains challenging. There is no larger any controversy about the necessity of integrating several types of features (molecular, morphological, ecological, etc.) and using various species delimiting criteria (phenetic, biological, phylogenic, etc.) but the question of the relative importance of these features and criteria is still a subject of debate. Nevertheless, whatever the type of features, the type of criteria, the analysis methods and the order in which they are studied, the sampling quality remains critically influential on the quality of the formulated taxonomic hypothesis as well as on the inferences about the structure and dynamics of biodiversity that are built after the initial processing.

Depending on the criteria applied, to define the boundaries between species in the genealogical network consists of identifying the phenetic differences between groups of organisms, deciding if the groups form distinct evolutionary lineages and testing the lack of gene exchanges between these lineages. All the criteria rely on estimations of the intra- and inter-species diversity based on samples of numerous specimens of each species and of numerous species within each studied set (taxon, community, etc.). Providing precise guidelines for the number of specimens per species to sample to ensure an accurate estimation of these diversities remains challenging. The sampling of a large vertebrate species endemic to an oceanic island does not require the same effort as that of a benthic mollusk that can be found in the whole Indo-pacific region. The sizes of data sets for integrative taxonomy are, however, nearing those of the data sets for "population genetics". At the interspecies level, the tendency is the same: within the group studied, the larger the number of species included in the analysis, the more representative the estimated interspecies diversity and the more robust the taxonomic hypotheses.

In this context, molecular analyses are regularly integrated into taxonomy methodologies; with for example "barcoding" typed projects. As in the case of specimens, the sampling of genetic markers within genomes should not be overlooked: one of the main criteria for the delimitation of species is their reproductive capacity (or incapacity), meaning gene exchanges and this criterion can only be tested if several independent genetic markers are analyzed. In many organisms, the identification and characterization of such markers (although easily accessible and variable at the species level) is often difficult because of a lack of knowledge about their genomes.

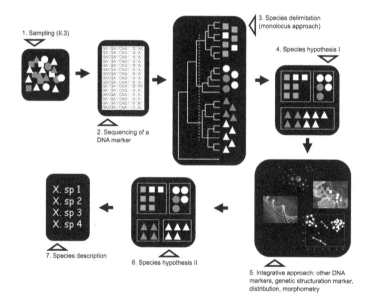

Figure 5.6. *Conceptual diagram of integrative systematics. Sampling (1) and sequencing a monolocus marker (2) leads to a first species delimitation hypothesis (3 and 4). The application of an integrative approach to the same specimens, including an NGS multilocus analysis, can give rise to a new species delimitation hypothesis (5) eventually leading to a new description of species (6), © Nicolas Puillandre. For a color version of this figure, see www.iste.co.uk/faurejoly/genomics.zip*

NGS approaches appear to be potential solutions to these issues. They enable the sequencing of many specimens as well as many genetic markers. Some of these technologies do not even require the definition of a target-gene, which means they do not require the genome to be *a priori* known. This is a great benefit for taxons whose genomes, diversity and taxonomy are not well-known. Furthermore, RAD-seq* methods ensure a coverage* of the genome that is much greater than "monolocus" methods. This therefore frees us from the problems linked to sequencing for one or only a few genetic markers within taxonomic approaches.

6

Characterizing Biodiversity

Before NGS, "DNA barcoding" was proposed as a means to identify species. It uses a short specific region of the genome where a variable DNA sequence serves as a specific signature that identifies a given species or taxonomic group [VAL 09]. DNA barcoding relies on the production of standardized genetic data (DNA barcodes) from reference specimens that are identified by taxonomists and recorded in collections, thus ensuring the application of biological nomenclatures.

In this context, the Barcode of Life (BoL) project develops this universal tool to assess species diversity in several fields, such as ecology, agronomy, forensics, etc. The project aims at accelerating the description of yet unknown biological diversity. It specifically focuses on Eukaryotes taxonomy and uses the mitochondrial gene cox1 for animals and a combination of chloroplastic (rbcL, matK, trnH-psbA) and nuclear (ITS) genes for plants. Other regions of genomic DNA, such as the genes that code for ribosomic RNA (rRNA) in bacteria, eukaryotes and archaea can also be used as barcodes, with different taxonomic resolutions. Several databases use these barcodes: the Ribosomal Database Project (RDP), Silva, and Protist Ribosomal Reference (PR2). Metabarcoding extends the barcoding approach: it aims to grasp the biodiversity within environmental samples (soil, water, digestion content, feces, etc.) by analyzing the barcodes in order to characterize simultaneously the DNA left by individuals from different species in the environment [TAB 12]. Metabarcoding uses various genetic standards depending on the required level of resolution (species level, infra- or supra-species) to characterize the degraded DNA extracted from the environment (see Chapter 8).

Figure 6.1. *Biodiversity inventories are established in all environments; top left, mangrove, © Gaelle Fornet/CNRS Guyane/CNRS Photothèque 20150001_0236; top right, coral reefs, © Thierry Perez/CNRS Photothèque 20140001_0417; bottom left, polar ice cores, © Joël Savarino/CNRS Photothèque 20070001_0047, bottom right, height of the canopy in a tropical zone, © Daniel Lachaise/CNRS Photothèque 20050001_0079*

6.1. Barcoding and metabarcoding in the NGS era

The first limitation of DNA barcoding and metabarcoding is the completion of the taxonomic databases used as references. The second relates to choosing genetic standards that offer enough resolution and cover the diversity of life. NGS technologies might overcome some of these limitations by making the characterization of a large number of markers easier, covering genomes more widely. This would especially increase the reliability and resolution of taxonomic assignments in biodiversity studies.

While *stricto sensu* DNA barcoding focused on improving the completion of reference databases and the relevance of taxonomic assignments, approaches that characterize environmental samples (metagenomics metabarcoding) took advantage of the developments allowed by NGS. Several laboratories use NGS to describe the diversity of life in systematics, phylogenetics and ecological studies. Each of these disciplines has its own specific issues and tools. However, the results that each one produces must be able to be utilized by the rest. For example, reference databases that relate genetic (barcodes) to taxonomic (type) information must be available to ecologists to characterize the diversity of ecosystems. Reciprocally, taxonomic and phylogenetic studies should quickly take into account new biodiversity compartments found in ecological studies. One major issue is, therefore, to design methodologies that are compatible with both systematics and ecology.

To reconcile the various properties necessary for a DNA barcode to be efficient both for taxonomic and ecological studies is indeed difficult [VAL 09]. The difficulty stems from the contradictory needs that arise from the standardization of reference data, the requirements of biodiversity studies, and the technical constraints. The large numbers of sequences provided by NGS make it possible to consider the development of multilocus barcodes (the combination of many short fragments) that more widely cover genomes and can be used in various types of study. The production of short fragments enable the characterization of degraded environmental DNA, and the large number of markers provides enough information to reach a good taxonomic resolution and a consistent phylogenic signal. Using multilocus barcodes enables a less reductionist approach for defining taxons, based on a larger number of independent features. However, gaps will probably remain in large taxonomic spectra, partly because of the lack of complete reference genomes for whole compartments of the tree of life. There is, furthermore, the issue of the homology of markers among compared organisms and thus the issue of the phylogenic depth of the obtained signal. These approaches produce, at least partly, non-standardized data whose exploitation for comparative studies is not straightforward (for example, a few million short sequences that represent a varying fraction of a genome may not be comparable across organisms). Testing hypotheses on the homology of markers can only be achieved on the basis of a taxonomic sampling that is relevant to the analyzed phylogenic depth. The problem of the completion of the reference database and thus that of the impact of missing data on the quality of inferences were already challenging when using only a few standardized markers. In this case, they are becoming more acute.

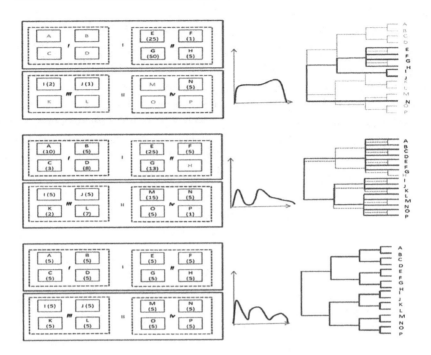

Figure 6.2. *Crucial importance of sampling and genetic markers used for taxonomic classification. On the left, the number of collected samples (number in brackets) for each taxon (rectangles labeled A to P), grouped into higher level taxonomic groups (dotted-line rectangles labeled f, ff, fff and fv, and rectangles in bold lines labeled I and II); in the center, distribution of genetic distance according to the chosen markers; on the right, inferred phylogeny of the various taxons. From top to bottom, three cases are compared: 1) partial sampling of the various taxons (left) and use of non-resolutive genetic markers: this leads to a distance distribution that is not multimodal (graph in the center) and to a partial representation of taxons and their relation in the phylogenic tree; 2) almost exhaustive sampling of the various taxons (on the left) and use of a resolutive monolocus maker for a specific taxonomic rank (e.g. species): this leads to a bimodal distance distribution (center) and to a better resolution of the phylogenic tree at the considered taxonomic level (for example, species); 3) sampling that covers all the taxons (left) with a NGS-based multilocus approach that enables the resolutions of various taxonomic levels (for example, species, genus...): the distribution of distances is multimodal and the phylogeny is robust at all taxonomic levels, © Sarah Samadi*

6.2. Adapting NGS approaches to environmental sample constraints and the knowledge of organisms available

DNA fragments are currently characterized after PCR amplification. This step introduces biases and errors and heavily constrains the definition of DNA barcodes to areas bounded by preserved sequences complementary to the primer's sequences. Several NGS-based approaches could overcome these constraints: (1) DNA capture enables us to bypass PCR to select, for later sequencing, target barcodes thanks to probes complementary to conserved regions within the barcode of interest; such methods can be applied for characterizing environmental samples but also for searching for homologous markers in distant organisms; (2) RAD-seq-based methods (section 6.2.1) could efficiently contribute to delimiting species and building phylogenies (this approach cannot be used to characterize degraded DNA from environmental samples); (3) massive direct sequencing enables the complete assembly of genomes of organelles (mitochondria and chloroplasts) and of multiple copy genes (ribosomic genes, etc.) that are already used to construct large-scale phylogenies (section 6.2.2). Applied to ecological samples, such metagenomic approach would enable the simultaneous characterization of taxonomic and functional diversities (section 6.2.3). Beyond the information provided by the targeted zones, this approach produces millions of short nucleic sequences which contain information that can be exploited for functional or taxonomic assignations as well as for phylogenies.

6.2.1. Fine-grained study of the diversity of the Chrysogorgia gorgone using RAD-seq

Knowing the taxonomic and phylogenic diversity of organisms is a prerequisite for environmental genomics projects. These projects require reference databases to properly relate environmental genomics data to labeled organisms. In this context, systematics faces challenges due to the magnitude of biodiversity and to the difficult task of transposing genomic methodologies to non-model organisms (constraints due to genome size and structure). How can we easily and robustly identify the members of taxons with an unknown genome? Below is the approach taken for the *Chrysogorgia* gorgone.

The *Chrysogorgia* genus, with about 60 described species, is one of the most diversified genus of octocorallia. It is a monophyletic genus that is widely distributed and whose morphological variability coincides with the genetic variations of the mitochondrial marker mtMutS. The diversity of this gene is however small, as morphotypes can differ for only one base. To validate the hypothesis according to which morpho-haplotypes correspond to species, the RAD-seq method [BAI 08] was used. It provided a high number of markers varying within and among haplotypes, which allowed us to define groups on the basis of effective gene flows.

Figure 6.3. *In the foreground, a yellow crinoid and on the right a colony of* Chrysogorgia *with a pink sea anemone in its lower branches (Atlantic Ocean), © Eric Pante. For a color version of this figure, see www.iste.co.uk/faurejoly/genomics.zip*

A prerequisite for this NGS approach was an appropriate sampling strategy that enabled us to test taxonomic hypotheses at various phylogenic depths. This strategy consisted of:

1) various morpho-haplotypes that were sampled in sympatry in order to assess the lack of gene flows at the genomic scale;

2) specimens of the same morpho-haplotype, sampled from remote locations or at various depths to highlight a potential cryptic diversity;

3) specimens of various mitochondrial clades to validate the phylogenetic signal of this marker.

The RAD-seq method generates a great number (several thousands) of informative SNP markers. Data analysis includes a bioinformatic analysis that sorts the sequences according to their quality, assembles the identical sequences of each individual and then compares them across individuals whilst evaluating the homology of markers. Measures of genetic divergence between *Chrysogorgia* individuals were in agreement with taxonomic decision made on the basis of mitochondrial data, as the intra-haplotypic divergence was clearly smaller than the inter-haplotypic divergence [PAN 15]. One of the major advantages of the RAD-seq method, provided it is associated with a relevant taxonomic sampling, is that it provides a large number of markers simultaneously and the comparative framework necessary for their interpretation. This method can be applied to many cases such as: fine-grained spatial characterization of population structures, detection of regions under natural selection pressure and also phylogenetics.

6.2.2. High-resolution megaphylogenics: the case of alpine flora

Over the past decade, the use of NGS-based genomic data in biodiversity research has experienced an unprecedented rise, especially in the context of:

1) the reconstruction of ever-larger phylogenies (megaphylogenies) that enable us to model the diversification of widely distributed clades and the history of the emergence of biodiversity hot spots;

2) the development of new approaches in the characterization of biodiversity patterns on the basis of environmental DNA from soil (metabarcoding technique). The example of the Phylo-Alps project shows the constitution of the reference genomic base of a complete biome, the Alpine Arc, to develop research on phylogeny, comparative genomics and metabarcoding in the region [ROQ 13].

The project is based on an effort to sample the whole Alpine flora systematically, using the revised pre-existing data of Flora Alpina as reference. The aim is to characterize 4,500 species and subspecies with 1 to 2 samples for each taxon and to constitute a reference herbarium with the sequenced plants. The NGS technology for this project is the Illumina HiSeq sequencer, which achieves a very low-coverage sequencing of the genome of each collected species (x 0.1). This sequencing enables to reconstruct the chloroplastic and mitochondrial genomes of the studied species and to extract the sequences of repeated nucleic genes (especially ribosomic genes) with assembly programs dedicated to this type of data. From the produced data, we can build megaphylogenies on the basis of genetic data that is identical across a large number of species.

Figure 6.4. *Megaphylogeny of the PhyloAlps project. Left, an example of megaphylogeny of 823 genera of plants of the alpine arc built with the supertree mixed method [ROQ 13]. On the right, examples of remarkable species and environments (top to bottom and left to right): prairie of Rhododendron ferrugineum (red flowers) of the Écrins National Parc; subalpine prairie with great diversity, specific of the Col du Lautaret; Androsace helvetica (plant in compact tufts) from the crest of the Galibier © S. Aubert;* Eritrichium nanum *(blue flowers) from the summit of the Alps, © S. Ibanez. For a color version of this figure, see www.iste.co.uk/faurejoly/ genomics.zip*

6.2.3. *Bacterial metabarcoding: intestinal microbiome in myrmecophagous mammals*

Myrmecophagous placentals (whose diet is made of more than 95% of ants and/or termites) are a typical example of evolutionary convergence [DEL 14]. They are animals such as aardvarks, anteaters, pangolins or aardwolves. In terms of adaptation, the intestinal microbiome (i.e. the set of genes of microorganisms) plays a major role and its evolution in mammals is influenced by the phylogeny of the host species and by their diet. The issue here is to test a major evolutionary hypothesis about the convergence of the microbiome due to the myrmecophagous diet, using an environmental genomics approach.

The analysis of fecal samples of myrmecophagous mamals and of related species, allows characterizing the taxonomic composition of the microbiote by Illumina metabarcode sequencing (*rrs* gene of bacterial communities). The taxonomic assignation of sequences and comparative analysis of bacterial communities are then performed. A second approach consists in sequencing the metagenomic DNA of bacterial communities of a representative subsample of these myrmecophagous species with Illumina sequencing. The sequences assembled from this metagenomic data are then functionally annotated by searching similarities with protein and functional data from reference databases. These two types of analysis will enable us to respectively characterize the taxonomic and functional compositions of the intestinal microbiomes of these species.

The first taxonomic part of the study enabled the analysis of almost 100 fecal samples with barcodes of the rrs gene (rRNA 16s) representing the bacterial diversity of myrmecophagous and related species. The representation, as a phylogenic network, of the distances between the various microbiomes reveals how samples can be grouped both by their phylogeny and by their diet. Therefore, two large groups of herbivores, corresponding to monogastric (capiraba, rhinoceros, elephant and zebra) and polygastric (giraffe, okapi, gazelle and sheep) can be distinguished by their microbiomes. This approach, furthermore, shows that myrmecophagous species can be grouped together as they have similar intestinal microbiomes despite their different phylogenic origins. In particular, the aardwolf, a hyena that specializes in eating termites, is grouped with the aardvark and anteaters rather than with the spotted hyena that is closer to other carnivorous species.

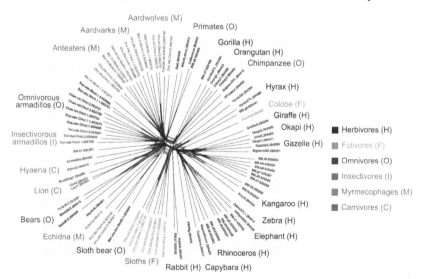

Figure 6.5. *Phylogenic network representing the relations between the diversity of intestinal microbiomes of various mammals that have different diets, © Frédéric Delsuc. For a color version of this figure, see www.iste.co.uk/faurejoly/genomics.zip*

6.3. Challenges to be met in analyzing high throughput biodiversity

To consolidate the use of NGS in the analysis of the dynamics of biodiversity, two main challenges need to be met. First of all, reference databases must be completed by increasing their taxonomic coverage. This not only involves standard barcodes, standardized regions (e.g. complete mitochondrial genomes) but also all the non-standardized genomic data (partial genomes, RAD-seq data, etc.), for which a good taxonomic coverage is necessary to solve homology issues. Therefore, the success of these new approaches is dependent on the development of methods for the taxonomic assignation and phylogenic inference that take into account sequencing errors and DNA degradation. This is especially pertinent in environmental and paleogenomic studies that make the best use of information coming from the available data regarding standard markers and/or partial genomic data.

7

Evolution and Adaptation
of Genes and Genomes

Fine-grained comprehension of the evolutionary history of phenotypic traits requires the identification of their genetic basis. The aim is to discover the genes or genomic structure involved in adaptive evolution, i.e. in response to natural selection. The panel of methodological tools that NGS provides enables the exploration and identification, in the whole genome, of important functional variants with their associated evolutionary or ecological changes, taking into account the confusing effects linked to the specific history of the studied populations.

This evolutionary genomic approach requires the integration of multiple sources of information deriving from molecular population genetics to phylogeny and formal genetics. The combination of these sources of information enables the identification of relevant functional variants. Until recently, these projects required significant human and technical investments due to the low throughput of the genotyping methods, of the inaccessibility of the unknown regions of the genome and the limits of multilocus statistical methods. Methodological developments have since revealed the importance of the multilocus dimension for the study of adaptive responses. Massive parallel sequencing solutions proposed by NGS overcome a barrier by providing a throughput which is sufficiently high to undertake the evolutionary genomic approach. This technique now enables us to address fundamental issues about the evolutionary process within populations, issues that were until now out of reach, or considered as pertaining to the field of theoretical biology. Indeed, it is now possible to understand the nature of functional changes and of the gene networks they involve, the overall distribution of mutations and their effects, the patterns of recombination and even the composite nature of genomes that undergo complex selection processes.

Figure 7.1. Gorgonia flabellum, *venus sea fan inhabiting the Antilles Islands, studied for their adaptive responses to hydrodynamic stress, © Jérôme Fournier/CNRS Photothèque 20130001_0051*

7.1. Discovery of recurrent selection signatures

Thanks to NGS technologies, which enable the sequencing of entire genomes on the basis of a single individual, it is possible to obtain precious information about intraspecific variability (polymorphism) via the heterozygosity of individuals. Because of recombination, an individual's particular genome reflects the genetic diversity of numerous ancestors and carries precious information about past processes including adaptive processes. Researchers have for instance tested the hypothesis of recurrent selection in primates, which means the hypothesis according to which the same loci would undergo independent adaptive changes in the various evolutionary lineages [ENA 10]. For example, when a beneficial variant spreads into a population, it alters the whole neighboring region of the genome, a process coined "selective sweep". This thus leads to a dramatic reduction of local variability along the chromosome segment. A classical selective test (HKA for example) relies on the detection of these regions exhibiting reduced diversity in comparison with the level of interspecific divergence that mirrors the local mutation rate. These tests have been adapted to entirely sequenced individual genomes and applied to a number of recently sequenced genomes of primates providing a list of loci that indicate potential signatures of natural selection. This list is mostly congruent with lists produced by other selection tests that focus on interspecies comparisons of the ratio between non-synonymous and synonymous (supposedly neutral) variability in the coding sequences of genes. This congruence thus confirms the validity of the test. The candidate genes, signaling a response to selection, especially include genes that are expressed in the cerebellum, spleen and testes. This suggests several adaptive changes related to defense mechanisms, to gametogenesis, and to the transcription and development of the forebrain. Hence, the same candidate genes are consistently selected in the evolutionary branches leading to the various primates, and not only in human lineage.

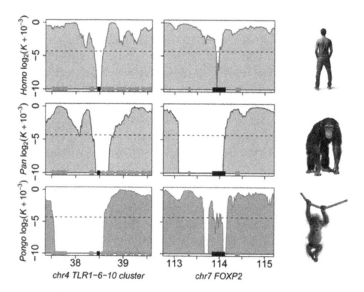

Figure 7.2. *Examples of candidate loci detected in several evolutionary lineages of primates. Selective sweeping was detected in the loci of Toll-like receptors 1, 6 and 10 (innate immune response; left) and of gene FOXP2 (right). Each figure covers 2 megabases (X-axis). The location of genes is given on this axis (black: candidate genes to selection, centered on lower variability zones, gray: other genes). Top: human (pink); middle: chimpanzee (blue); bottom: orangutan (green). This study challenges some interpretations about specifically human adaptation as, for example, the so-called language gene FOXP2 that is found in the common ancestor, © David Enard. Frantz Depaulis and Hugues Roest Crolluis, from [ENA 10]. For a color version of this figure, see www.iste.co.uk/faurejoly/genomics.zip*

7.2. Target genome sequencing methods

The discovery of genetic and genomic variations involved in adaptive evolution gains from the implementation of the whole range of current NGS technologies using genomic DNA and transcriptomic RNA material. Although NGS technologies significantly reduce the cost of sequencing each pair of bases on a base pair basis, each experiment nevertheless involves substantial investments in sequence production and, above all, in the analysis of the generated sequences. Furthermore, the large size of many eukaryote genomes imposes per individual costs that can be limiting. Optimal strategies therefore often use a sub-sampling of genome complexity, depending on the genomic region of interests and on the available material ("reduced representation libraries"). The various methods for this sub-sampling are briefly introduced below:

– The *RAD-seq* enables a significant sub-sampling of the genome around enzymatic restriction sites, with strong possibilities of multiple sample pooling* with moderate costs. It is useful for population genetic mapping, for genomic scans of association as well as for phylogenetic or phylogeographic multilocus analysis on recent clades for which the classical methods often can't achieve the required resolution. At deeper phylogenetic levels, obtaining preserved markers in sufficient amounts is often difficult. Current challenges include the efficient use of RAD-seq without access to a reference genome*.

– The *sequence-capture of methods* enable the optimization of the sequencing effort by specific enrichment of large or very large genome portions before sequencing them with high throughput. Such enrichment relies on hybridization of DNA or RNA with probes elaborated from a reference sequence. These methods are an alternative to the re-sequencing of entire genomes as they enable approaches that focus on specific target regions of the genome (chromosome segments, family of genes) and are also large-scale. A major benefit of this experimental strategy is to enable significant sequencing coverage on genomic regions whose size and complexity prevent the amplification with traditional sequencing. This is the reason why targeted re-sequencing is often used when searching for nucleotidic variants, but also structural variants of the CNV (copy number variation) type, which can be detected, for example, by methods accounting for the depth of the coverage. The benefits of these approaches are gradually diminishing, however, as the costs of sequencing itself drops in comparison to the costs of preprocessing, which remain quite expensive. They, nevertheless, remain very relevant for the analysis of coding regions of genomes that are especially rich in transposable elements or for the study of taxons whose genomes are too large. The current challenging issues include the capture of sequences that are somewhat divergent from reference sequences as well as the optimization of the multiplexing procedure or "pooling" before capture, in an attempt to reduce the costs of preprocessing.

– The *transcriptome* as a sub-sampling method focuses on expressed sequences, which is useful in an approach based on the comparison of genetically distant taxons, for example to compare molecular polymorphism and divergence in non-model groups. It is also used to produce a reference to annotate a genome or for other applications (for example sequence captures). However, since the sequencing is quantitative, this type of methods, such as for example RNAseq, is mainly used to search and study the expression of sequences (genes, mRNA, transcripts) that exhibit expression or splicing variations. RNAseq data is therefore used often to compare transcriptional variations across various phenotypes, and there are numerous analysis pipelines* therefore. Current challenges include transcriptome assembly from short reads NGS technology and the complexity of data sets, which require a lot substantial computing power as soon as the number of biological or technical replicates increases.

– *Re-sequencing genomes* gives access to the complete description of genetic variations and is becoming affordable for species with small or medium-sized genomes. Since the sequence fraction obtained is not based on a reference, the data produced is independent of the genetic distance from known taxons, which is actually useful for some comparative approaches. Furthermore, the preparation of libraries does not require additional steps and their machine processing is a routine operation of sequencing centers. All this ensures that complications are kept to a minimum and thus a significant amount of time is often saved.

The challenging issues that this approach must tackle now consists of optimizing the trade-off between sequencing depth and genotyping quality because for some applications like genetic mapping, a relatively superficial sequencing is sufficient because a less extensive allelic diversity is required. To bypass the complete re-sequencing of individual genomes to produce polymorphism data within a population, one can sequence a blind mix of individuals of a single population [SCH 14]. This strategy, called Pool-Seq, lowers the cost of producing polymorphism data and, under certain conditions (high number of pooled individuals), minimizes the variance of allelic frequency estimators. This approach seems especially efficient to identify genomic regions that significantly differ between populations of different environments. Another alternative to complete re-sequencing consists of sequencing only a targeted zone to increase the coverage in these regions (see, at the beginning of this chapter, the example of recurrent selection in primates).

Figure 7.3. *The pea aphid, a well studied species for the processes of speciation on host races,* © *Carole Smadja*

7.2.1. *Target sequencing, host adaptation and specialization in the pea aphid*

A study on the genetic basis of the specialization to the host plant was carried out by capturing candidate genes in the pea aphid *Acyrthosiphon pisum* (Figure 7.3). Several host races of pea aphid exist in sympatry, being highly specialized to various types of leguminous plants and this ecological specialization is the key element of the reduction of gene flows between biotypes. Since the choice of the host plant involves mechanisms of gustatory and olfactory recognition, the chemical sense genes, who are organized in large multigenic families in the aphid's genome, are good candidates for the study of adaptation in this insect species. Re-sequencing approaches that target these gene families have been developed to detect genetic changes at the origin of the divergence between host races. Two major results are drawn from these studies [SMA 12]:

1) the identification of a dozen olfactory and gustatory genes that are candidates to the adaptation to the host plant as they exhibit the signature of selection on biotype divergence;

2) evidence that some candidate gene families, especially the olfactory receptors and the odorant binding proteins, exhibit a divergence in copy numbers between host races, which suggests that segment duplication plays a role in adaptive diversification.

7.2.2. Target sequencing and hybridization studies on the domestic mouse

This study focuses on the genetic basis of sexual isolation and of reinforcement in the domestic mouse *Mus musculus,* by re-sequencing of entire exomes. The two European sub-species of the domestic mouse, *M. m. musculus and M. m. domesticus*, have diverged in allopatry for 0.5 million years before they met in Europe where they formed a hybrid zone in which behavioral barriers to gene flow seem to have evolved as a response to hybrids selective disadvantage (suggesting a reinforcing phenomenon). A strategy of re-sequencing the entire exome of mice was implemented to discover the genomic regions involved in this sexual isolation. Comparing the levels of polymorphism in allopatric populations versus populations that are in close contact enabled as a first step to identify genes that undergo selection pressure due to this putative reinforcement phenomenon in the hybrid zone. Furthermore, analyzing the levels of coverage of the various regions of the exome enabled researchers to identify, in the various samples, potential variations in copy numbers (CNV) that could affect some candidate genes such as vomeronasal receptors, olfactory receptors, genes that code for the Major Urinary Proteins or for the major histocompatibility complex. The involvement of such genes is relevant in mice because their reproductive behavior mostly depends on chemical signals between sexual partners. These genetic factors, as a whole, could therefore be the basis of adaptive behavioral divergence in mice [MA 15].

Figure 7.4. *Domestic mouse,* Mus musculus, © *Carole Smadja*

7.2.3. *Comparative transcriptomics in nickel hyperaccumulator plants*

This technique to study the transcriptome has been applied to molecular research on plants that can detoxify soils especially nickel hyperaccumulator plants. There are about 500 plant species (0.2% of angiosperms species) that are able to accumulate signification concentration of metals in their parts above ground. Remarkably, 400 of these species (grouped in about 40 families) can accumulate more than 0.1% (dry mass) of nickel in their parts above ground [KRA 10]. The hyperaccumulator species live in ultramafic or serpentine soils that are rich in nickel, mainly in tropical zones (e.g. Cuba, New Caledonia), but also in temperate climate regions such as Europe. These plants are, nowadays, a significant asset for the development of sustainable biotechnologies such as the phytoremediation of polluted soils, phytomining and green chemistry. However, the molecular mechanisms involved in nickel accumulation are still hardly known because hyperaccumulating species have not benefited from the most advanced research and tools developed for model species or species that interest the agribusiness field. To identify the genes involved in nickel hyperaccumulation in plants, NGS technologies are used to obtain genomic sequences of these plants and to study their gene expression.

In a first step, the reference transcriptome of the *Psychotria gabriellae* species (also known as *P. douarrei)*, an endemic Rubiaceae of New-Caledonia that can accumulate up to 4% of nickel in its leaves, was sequenced. The *de novo* assembly of reads produced by GS-FLX (Roche 454) provided 30,000 contigs and constituted the first database of expressed gene sequences of this species. Genes that code for membrane proteins able to carry metals were then identified and cloned. One of the identified transporters, homologous to the ferroportin that carries iron in mammal cells, is able to carry nickel in plants. Furthermore, this transporter is less expressed in the close species *Psychotria semperflorens* which, although it lives in sympatry with *P. gabriellae*, does not accumulate nickel. These results, therefore, suggest that this transporter from the ferroportin family plays a role in the hyperaccumulation of nickel in *P. gabriellae*. This type of study demonstrates the value of biodiversity and non-model species and opens perspectives regarding the potentially important use of identified molecular mechanisms in remediation processes or in ecosystemic services linked to land-use changes.

Figure 7.5. Psychotria semperflorens *(left) and* P. gabriellae *(right) observed in sympatry in the rainforest with ultramafic soil in New-Caledonia. In this context,* P. gabriellae *accumulates more than 1% of nickel in its leaves as shown by the pink color exhibited in the presence of dimethylglyoxime (inserted picture), whereas* P. semperflorens *accumulates 100 times less. The transcriptomes of these phylogenietcally close species can be compared in order to identify genes whose expression is linked to the accumulation of nickel,* © Sylvain Merlot. *For a color version of this figure, see www.iste.co.uk/ faurejoly/genomics.zip*

7.3. Characterization of a reference genome for adaptation studies

The availability of a reference genome is a major asset for most of the approaches described above. To document structural variations between genomes, the detection of synteny breaks (points where the order of loci is not preserved between taxa) can for example be achieved by analyzing differences in mapping locations on a reference genome. The currently challenging issues include the availability of assembled genomes and the quality of their assembly, which may affect, for instance, the power to detect type 1 errors (false positives). In comparative micro-evolutionary studies, a reference genome allows the use of sliding window approaches that exploit the whole genomic continuity along the conserved fragments (scaffolds or superscaffolds) and therefore increase the power to detect relevant signals of population differentiation. NGS technologies open the possibility to generate complete genomes of eukaryotes of a few hundred megabases, with a deep sequencing (100X or more by Illumina 100 bp) and to use libraries with varying fragment sizes to optimize the assembly. Other types of next-generation sequencing methodologies that, like PacBio, do not rely on the amplification of DNA fragments enable us to produce much longer fragments. Although they generate a much higher error rate, they enable, in combination with Illumina sequencing, to significantly increase the assembly efficiency. Assembly is actually facilitated when additional genomic resources are available and even more when the material on which it is based is homozygous. Beyond the requirement for deep coverage, which raises sequencing costs, current challenging issues for the study of non-model organisms lie in assembly optimization and especially in the bioinformatic treatment of heterozygoty and of strong allelic differentiation.

7.3.1. NGS strategies to study the adaptation of a non-model species, the tropical butterfly Heliconius

The characterization of a reference genome was achieved for *Heliconius* butterflies and enabled the identification of the origins of evolutionary convergences between various species thanks to a population re-sequencing. The neotropical genus *Heliconius* exhibits a spectacular radiation-diversification due to mimetic behavior [JOR 06]; an increasing number of teams work on the genetic determinants of their convergence. To produce a complete reference genome, these teams informally linked to each other in an international consortium (www.heliconius.org) to take advantage of the NGS advances from the very beginning of their advent and thus managed to sequence the genome of the *H. melpomene* species (295 Mb haploid genome, [HEL 12]).

The 454 Roche pyrosequencing, which has a lower throughput than Illumina but provides longer reads (400 bp), was initially chosen for the first assembly (12M reads on a consanguineous individual). Paired-end Illumina sequencing, and especially the sequencing strategy called "mate-pair", which uses circularization of long fragments (3 kb and 5 kb), was then used to associate the contigs separated by sequence repeats into scaffolds (8 M read pairs). The development of 100 bp reads by Illumina HiSeq enabled the production of 42 M additional read pairs, which improved the average coverage and enabled the correction of pyrosequencing-inherent errors. The final assembly is, therefore, made of ~269 Mb (91% of the genome, 3800 scaffolds, N50=277 kb). Lastly, individuals originating from a Medelian crossbreeding between the consanguineous lineage and a divergent lineage were genotyped by Illumina RAD-Seq to produce a genetic mapping that enabled us to arrange and orient on each chromosome the genomic scaffolds containing RAD markers (super assembly of 83% of the genome, 1,273 superscaffolds, N50=400 kb) (www.butterflygenome.org).

The sequencing technologies and analytical methods, in constant development, have drastically changed during the project. The sequencing strategy was, therefore, continuously updated to keep up with the increasing power of technologies as well as adapted to the assembly issues raised in the course of the project. Most of these issues were related to the heterozygosity of the genome, a factor that makes the assembly difficult, and to its haplotypic structure, which generates an artefact duplication of strongly heterozygous segments. The sequencing of genomic libraries of variable sizes and the implementation of *ad hoc* bioinformatic strategies partially solved these issues [HEL 12]. The genome produced, as such, is very useful for a wide range of applications, from structural genomics to adaptive processes. It will nevertheless be constantly improved [DAV 16, v2] notably with the sequencing of siblings (from mapping families) and the use of PacBio sequencing, which will together enable a better phasing as well as the correction of assembly errors.

To understand the structure of genetic variations across taxons and across mimetically convergent groups (Figure 7.6), the assembled genome was used as a reference to align population re-sequencing reads (Illumina HiSeq) on the localized genomic scaffolds, especially on the loci controlling color specification.

The continuity and genomic position of scaffolds enable sliding window (10 to 50 kb) analyses that reveal the variations of polymorphism patterns across multiple variable positions and even allow the analysis of variations of taxons that are not very different. Here, the genomic intervals that control color exhibit a phylogenic topology that is clearly structured by phenotypes whereas the adjacent regions as well as the genome as a whole are structured by taxons and geography. Other analyses based on the sharing of derived SNPs (ABBA-BABA) [HEL 12] reveal a signal of introgression between mimetic taxa, specifically associated with wing-patterning loci. These loci therefore tell us an evolutionary history that is very different from that of the population in which they are found. Here, mimetic convergence has evolved through the exchange of color alleles between species (thus, enabling an "instantaneous" resemblance) and not, as was thought before, through the parallel evolution of similar phenotypes in each taxon.

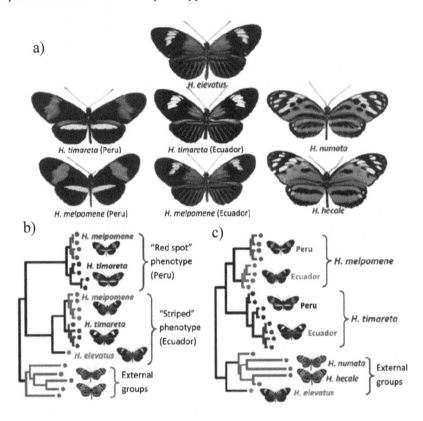

Figure 7.6. *Genetic origin of mimetic convergence in* Heliconius. *a) Three distinct mimetic groups that show the convergence between* H. melpomene *and* H. timareta *in a valley of the Andes in Peru (left "red spot" phenotype), between other races of that same species as well as* H. elevatus *in equatorial Amazon rainforest (middle column "striped" phenotype) and between* H. numata *and* H. hecale *in these same regions [JOR 06]. b) "Mimetic tree", phylogenic topology that precisely corresponds to the genomic interval of 50 kb that contains the locus of mimetic variation, where individuals are grouped by phenotypes and not by species. c) "Tree of species", phylogenic topology corresponding to most of the genome, where the taxons are grouped by species. B and C from [HEL 12], threshold = 1% divergence, © Mathieu Joron. For a color version of this figure, see www.iste.co.uk/faurejoly/ genomics.zip*

The *de novo* assembly of a genome, together with population re-sequencing thus enabled us to exploit the whole genomic variation and to reveal adaptive introgression of the color alleles, which led to an unexpected scenario to explain a phenotype convergence phenomenon between species.

The use of molecular polymorphism data to infer adaptive processes requires taking into account the demographic history of populations. Unfortunately, demographic inferences, usually require intensive numerical simulations for coalescence-based models and are typically computationally intensive; therefore they are poorly-suited to the massive amounts of data generated by high throughput sequencing. When appropriate, a solution to the computational limits may reside in the use of simple statistics and a detailed study of the regions with the most significant statistics, or previously identified regions [BUR 15, STA 15]. This however involves some risks (false negative, false positive), therefore the optimal exploitation of data sets requires the development of innovative methods that allow both a great flexibility of underlying models and a significant numerical efficacy [GOM 11, GUI 14].

As a conclusion on these diverse methods to study the evolution and adaptation of genes and genomes, we note that the use of NGS is strongly limited by access to sequencing platforms in a flexible framework and at reasonable costs. Some countries benefit from efficient university-based platforms that provide the local scientific community with direct access to technologies and rapid updates. By contrast, teams that do not have access to sequencing capacities that measure up to the present scenario must often use commercial biotech services which are not usually leading the field in terms of technical developments, only propose standardized protocols, and lack flexibility with regards to the research objectives in environmental genomics. Consequently, a real integration of, on the one hand, the formulation of relevant scientific questions and, on the other hand, technological, methodological or bioinformatic innovation would bring about the synergy necessary to propel advances in this field.

7.4. Conclusion and perspectives

Among the many scientific and technical challenges raised by the study of the evolutionary history of genes and genomes, we can highlight three main ones which should receive special attention towards efficient and cost-effective analysis of adaptation from genomic data:

1) statistical detection of signatures of selection on the basis of population NGS data: current tools are partly unusable on the whole genome because of the scale of genetic variation and of the complexity of its structure at the genome scale;

2) the efficacy of alignment* to a reference genome or of sequence capture in the context of strong natural intra- or interspecific variability: some strongly divergent genomic regions are resistant to the use of NGS approaches and their analysis can introduce significant biases that may lead to misinterpretation of adaptive evolution at the molecular level; developing a better documentation about variations in these regions of the genome as well as a better understanding of their effects is crucial;

3) the development of efficient bioinformatic pipelines for *de-novo* assembly of genomes in non-model species: many inference tools depend on the availability of reference genomes that are by definition not available for non-model species; the challenges are about the complexity of data sets as well as the question of heterozygozity, polyploidy and structural variability at diverse scales (indel, rearrangement, etc.).

As a conclusion, studies that combine several levels of analysis involving NGS lead to the discovery of genetic variants that enable to test adaptation scenarios that, until recently, were mere speculations. However, since the throughput of sequencing often exceeds the rate of development of the relevant analytical tools, a large part of the most emblematic studies involves important analytical innovations for genomic data [BAI 08]. Such innovations often rest upon three pillars: expertise about population biology, sequencing platforms, bioinformatic expertise and infrastructures. For the institutions that develop them, these pillars constitute springboards to place themselves at the forefront of the community and reinforce their international leading status. In Europe, national structures show various levels of progress towards a synergy between the formulation of fundamental questions about adaptive processes and the capacities of data production and analysis. The European community must, therefore, make specific efforts to support ambitious research on environmental genomics and promote assemblies of consortia with complementary expertise in these three disciplines.

8

Degraded and Paleogenomic DNA

DNA is a molecule that can survive in tissues even after the death of an organism. DNA can be extracted from most ancient organic remains such as bones, skin, teeth, appendages, feces, wood, seeds, shells, etc. Regardless of time, the state of such ancient DNA depends on the characteristics of the environment. In particular, the processes of preservation/degradation of DNA are influenced by the pH, temperature, humidity, pressure and DNA substrate adsorption. Hot deserts and caves, where the climatic fluctuations are limited, as well as permafrost and glaciers are prime locations where potential sources of ancient or degraded DNA can be found. A significant amount of paleogenetic information is also stored in museum samples, especially in the natural history museums of many countries (see Chapter 1).

Although DNA can be preserved throughout time in degraded environments or matrices (preserved food, leather, etc.), only a very small amount remains. Its substantial degradation into fragments of approximately 150 base pairs is worsened by chemical modification due to hydrolysis and oxidation. Furthermore, this ancient DNA is often extracted from samples together with inhibitors of the PCR amplification reaction that undermine subsequent analysis. All these factors result in the fact that ancient or degraded DNA is often subject to exogenous contamination by contemporaneous DNA, which is favored by PCR amplification.

The major issue that paleogeneticists and paleogenomists must address consists of ensuring that the sequences produced from ancient or degraded DNA are authentic and of verifying that they are not altered either by contamination or artefactual mutations (especially purine oxidation). To do so, researchers develop specific techniques and methods [ORL 15]. The challenge really consists of guaranteeing the authenticity of the studied DNA, which is required to investigate the evolutionary history of species on the basis of DNA fragments extracted from bones found, for example, in stratigraphic sediment layers. Therefore, target-focused expertise and dedicated machines within customized analysis platforms are required. In particular, paleogenomics laboratories must have confined dedicated rooms (pressurized and UV-irradiated) to limit the propagation of contemporaneous ambient DNA.

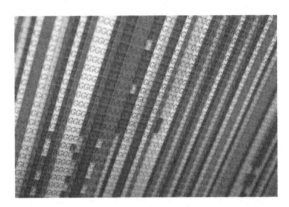

Figure 8.1. *DNA sequence analysis. Each color corresponds to a DNA base, © Vincent Moncorgé. For a color version of this figure, see www.iste.co.uk/ faurejoly/genomics.zip*

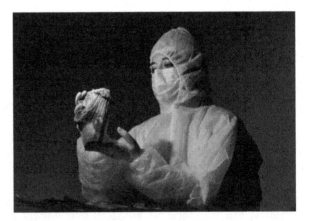

Figure 8.2. *Paleogenetic analysis of bone samples. Sampling is achieved with a dedicated platform, requiring strict controls to limit the propagation of contemporaneous ambient DNA © Vincent Moncorgé*

NGS technologies open up new opportunities for the analysis of this ancient or degraded DNA: they give access to the massive sequencing of small DNA molecules that are naturally predominant in the samples [HAG 15]. They enable the analysis of thousands of amplicons* that are obtained by targeted PCR amplification. Their comparison makes it possible to identify and characterize artefactual mutations in contrast with biological mutations and therefore authentic sequences can be produced.

NGS technologies have given access to the sequencing of the nuclear genome that is widely prevalent (in pure DNA) in the degraded eukaryote remains, but that is difficult to analyze by targeted PCR because of the dilution of DNA sequences (in terms of genomic copy numbers) in samples. These technologies also allow us, for some samples, to bypass the fastidious targeted PCR amplification step. Lastly, they enable the establishment of perennial libraries of degraded DNA produced on the basis of samples with very small amounts of endogenous DNA that are difficult and even sometimes impossible to amplify.

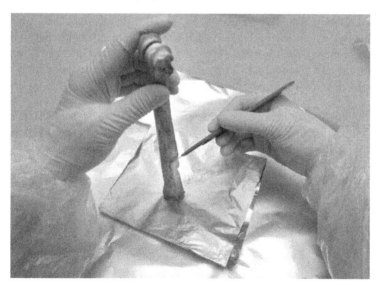

Figure 8.3. *Biological sampling by scraping a bone to extract DNA, © Vincent Moncorgé*

Technical and methodological advances on ancient or degraded DNA analysis are only the tip of the iceberg as conceptual developments are associated with the continuous technical progress, which has been especially significant over the recent years. The ability to extract DNA from degraded or ancient substrates provides population geneticists, systematicians and conservation biologists with the possibility of taking into account data coming from past individuals or populations. They then compare the sequences of ancient DNA to the corpus of contemporaneous data and therefore develop a diachronic outlook, i.e. a perspective on the evolution throughout times of some genetic traits [GEI 13].

8.1. Effects of domestication: the example of the origin and evolution of dogs

Domestication by humans of plant or animal species induces a fast evolutionary process that involves genetic, phenotypic and behavioral modifications, thus modulating biological diversity. The results of these processes can be observed nowadays on the various domestic animal breeds. However, ancient domestication processes cannot be studied on the sole basis of contemporaneous observations of genetic diversity. Thanks to the production of genetic/genomic diachronic data, paleogenetics has opened access to the genetic diversity of the past, to its variations during the various steps of domestication and to the understanding of species adaptation with regards to their relationships with humans.

The dog (*Canis lupus familiaris*) was the first animal to be domesticated, by the hunter-gatherers of the Upper Paleolithic (30,000 to 15,000 BP). More recently, after 300 years of intensive selection, the canine population is nowadays fragmented into 350 breeds that are phenotypically well characterized. This selection being very recent, current genetic data does not enable us:

– to access the phenotypes of the first domesticated dogs;

– to access the genetic variability that underlies these phenotypes, nor their diffusion through time and space;

– to understand the relationships between ancient genetic and phenotypic diversity and the diversity of contemporaneous races.

Thus, although we know that dogs could only be domesticated in areas where its wild ancestor the wolf (*Canis lupus*) lived, the origin and number of dog domestication occurrences have been the subject of debate for a long time [LAR 12].

Figure 8.4. *Wolfdog from Saarloos, © Benoit Rocher Wikimedia commons*

To address this issue, ancient dog and wolf bones from 41 archeological sites across Eurasia have undergone paleogenomic analysis. The analyses were carried out in a manner that was careful to reduce the risks of contamination to a minimum, during each step from sampling to sequencing. Only the authenticated sequences were taken into account. In the first phase, 68 authentic mitochondrial DNA sequences (D-loop) were produced and opened access to the genetic diversity of ancient dogs [PIO 10]. This data was compared to morphometric data. The results showed that dog domestication occurred in at least two separate regions during the Paleolithic, Asia and Western Europe (Figure 8.5); then later, during the Neolithic period, in the Middle East. This suggests that several populations of wolves generated the current dogs and that the first dogs were probably characterized by a high genetic and phenotypic variability.

A second approach consisted of producing data about the color of the first dogs (Figure 8.5), because color variation is one of the first effects of domestication. The analysis of the variants of two nuclear genes coding for the color of the coat, *Mc1r* (Melanocortin 1 Receptor) and *CBD103* (canine beta-defensin), showed that color started to vary as soon as the Mesolithic [OLL 13]. This paleogenomic approach enabled us to unveil not only the diversity coming from the genetic pool of the populations of wolves that were captured during the process of domestication but also the appearance of new variants due to the domestication-induced relaxation of natural selection pressures. These discoveries were possible thanks to the comparison of the already available genomic data of contemporaneous dogs (full genome of the boxer, annotated sequences, SNPs, etc.) with the data produced for ancient dogs and wolves.

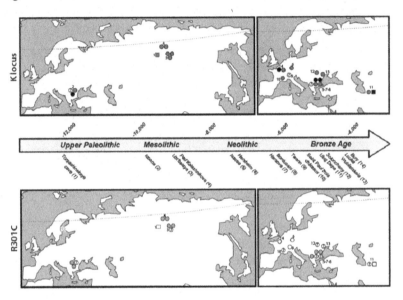

Figure 8.5. *Distribution of mitochondrial haplogroups and K^B (CBD103) and R301C (MC1R) alleles before and after Neolithic. Presence of the K^B allele (black), lack of K^B allele (dark grey), presence of the R301C allele (light grey), lack of R301C allele (white), undetermined color (question mark). Squares: wolves, circles: dogs, © Morgane Ollivier*

8.2. Biology of conservation

The paleogenomic approach also interests the discipline of conservation biology. Many species that are extinct or facing extinction benefit from re-introduction programs that, thanks to recent molecular techniques, allow us to better optimize operational strategies and practices.

The European sturgeon, for example, is a taxon that carries a strong symbolic and patrimonial value, because it is a fish that produces caviar as well as a great migratory animal. Having disappeared from the rivers of Western Europe during the 20th Century because of overfishing by humans and habitat modification, especially because of dams, sturgeons are a critically endangered species registered on the red list of the International Union for Conservation of Nature (IUCN) and the object of a national action plan in France within the framework of the national biodiversity strategy (NBS). Sturgeons are, in fact, spread over a complex set of species of which 17 of the 25 caviar-producing species belong to the *Acipenser* genus. Beyond legal regulations on caviar importations and controls on its origin (a black market that represents millions of Euros), an action plan in favor of its re-introduction in French rivers is being carried out. Paleogenomics helped this program by determining which species were present in the largest French rivers, on the basis of dermic scales extracted from archaeological sites, to suggest the species that are potentially best-suited to re-introduction in these ecosystems. Samples from the Rhône, dating from 2500 years before the present and more recent samples (from taxidermized specimens) have been analyzed [CHA 16]. Comparisons with libraries of other species of the same family showed that the European species *Acipenser sturio* was the only species present in this area, and so this speices was chosen for the national re-introduction action.

Figure 8.6. *Eurasian sturgeon* Acipenser ruthenus, *common species of Central Europe, © Vltava Wikimedia commons*

Other symbolic examples with high patrimonial impact include conservation programs for big mammals like tigers. A large amount of money has been invested in the conservation of these species, even though their ranges and habitats are nowadays very limited and subject to destruction, and the species are subject to recurrent poaching. In Asia, tigers now occupy only 10% of their initial range and molecular data shows a collapse of genetic diversity linked to a steep decline in population numbers during the Pleistocene. While nine subspecies were formerly identified, a recent analysis that combines morphological, molecular and ecological data formally characterizes only two subspecies. These correspond to two different ecotypes, one from continental Asia and the other in the south of the range in islands of the Sunda continental shelf, comprising Malaysia, Sumatra and the Philippines. The differentiation of these two ecotypes is probably due to a relative genetic isolation in addition to local adaptations related for example to the colder climate of the north of the range of the continental ecotype [WIL 15]. The international conservation programs that are currently being implemented and aim to double the number of tigers by 2020 should, therefore, take this genetic differentiation into account and promote three managing units, two on the continent and one for the islands, to restore the global genetic diversity of the species and reconstitute the local populations. This typically exemplifies the role of academic research and of the study of sequencing data in biodiversity conservation programs with societal stakes of significant economical and/or patrimonial impact.

Figure 8.7. *A tiger, one of the most emblematic species of conservation action,* © *Pierre Ferrière*

8.3. Molecular identification of manufactured goods

Trading of endangered species causes great losses to biodiversity. Every year, millions of animals threatened by extinction are illegally killed or captured to provide for trophy hunts, private zoo collections, decorative objects, traditional medicine or human consumption. Many methods to identify species in manufactured biological products have been developed. Until recently, they only relied on morphological studies, protein analyses, chemical analyses, etc. However, such methods have now become outdated since the recent developments in molecular biology and in paleogenetics have become well-suited to the analysis of degraded DNA in processed goods [TEL 05]. This new approach, associated with the recent development of barcoding and sequencing methods enables better fraud detection and food quality assessment.

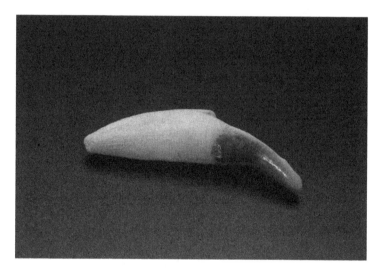

Figure 8.8. *Fossil dog tooth sampled for paleogenomic analysis, © Vincent Moncorgé*

The molecular identification of the origin of the species in processed goods like leather, fur and all goods manufactured from animal skins is especially important. This could be for trading control (to curb the illegal trade of endangered species) or for more fundamental research about the customs and activities of human populations during their evolutionary history. The taxonomic determination of skins is particularly difficult due to the treatments they undergo, such as tanning and coloring, which make the morphological determination difficult while accelerating DNA degradation. Specific DNA extraction methods have thus been set up to reduce the amount of extracted DNA inhibitors coming from the tanning and coloring treatments [MER 14]. These particular silica-based extraction methods that do not use any additional purification or animal serum for PCR enable us to produce reliable barcodes to reconstruct the origin of the species of the manufactured goods.

Such a method can also be used to re-create unique objects for which little information exists about the material from which they were made. The Museum of Music in Paris, for example, endeavored to rediscover the music of the past and to build a reproduction of an Erard piano (created in 1802) of which there exist only 13 specimens in the world and on which Beethoven, among others, played. The identification of the leather that covers the hammers was achieved through ancient and degraded DNA extraction methods and displayed a combination of ovine and bovine leathers [MER 16]. Leather from these two species was thus used to cover the hammers of the recently built reproduction piano.

8.4. Studying the human being: from evolution to identification

Human remains are also studied with paleogenetic and paleogenomic analyses. They necessarily imply even tighter precaution measures to avoid contamination by contemporaneous human DNA. Very varied research domains can therefore be approached. In the field of human evolution, a significant research effort has been invested in sequencing the genome of the Neanderthal, and it has been available since 2010 (www.eva.mpg. de/neandertal). The main benefit of this work is the production of a reference about human lineage, the *Homo* genus, as Neanderthal man is a sibling group of modern humans. By comparing these two siblings, we can now highlight the specificities of our species, *Homo sapiens* [ORL 06].

The cohabitation of Neanderthal with modern humans in Europe 35,000 years ago has fueled many debates in the literature about the possibility and degrees of hybridization between the two species. Over the past few years, several paleogenomic studies have prompted discussion thanks to new data and the subject still draws a lot of attention.

The improvement of degraded DNA analysis methods has societal consequences not only in the fields of environmental sciences and biodiversity studies, but also in all the questions related to species and individual identification. That is why in the fields of forensic science and legal medicine, genetic research based on the most recent DNA (and more specifically degraded DNA) sequence identification methods plays an increasingly crucial role.

In this context, genetic analysis essentially relies on new sequencing methods that enable the analysis of degraded DNA samples and on available genomic libraries. For example, the human genetic libraries have previously been analyzed using 13 micro-satellite markers presenting each between 8 and 13 known variations as well as a sexual marker to determine the presence of a chromosome X or Y. More recently, the use of about 50 SNPs, from the 3 million in that the human genome, is an improvement as it enables the detection of fragments of a smaller size than satellite or micro-satellite markers (miniSTR) and, above all, it is cheaper thanks to NGS. Therefore, thanks to cellular fragments collected on a crime scene (blood, hair, skin, mouth cells, saliva, sperm), identifying an individual via comparison to reference genomic libraries is now possible.

In the field of history many studies have also used paleogenetics to identify the remains of historical figures like Tsar Nicholas II, the American gangster Jesse James, etc. New fields of investigation are therefore opened up by this approach.

Figure 8.9. *Storing samples for paleogenomic analysis,* © *Vincent Moncorgé*

8.5. Conclusion and perspectives

Paleogeneticists are in some way the archaeologists of DNA. Thanks to the development of new techniques, new methods and new concepts, paleogenomics has gained considerable momentum, opening up access to the genetic heritage of many species and populations that are now extinct.

What the analysis of ancient or degraded DNA teaches us, globally, about the structure and dynamics of biodiversity spans several temporal dimensions. The analysis of the phylogeographic dynamics of species threatened by extinction or, on the contrary, those that risk invading is greatly eased by the access to non-invasive or environmental degraded material that does not require the direct sampling of taxons, the interest of which is sometimes elusive. Focusing on a more remote past, diachronic biogeography, referring to the evolution throughout time of modern and extinct species, sheds new light on their current dynamics.

Paleogenomics is currently achieving a significant breakthrough in population genetics, as it enables us to trace the phylogeography of species throughout time and thus to determine the dynamics of the structuring of populations, especially as a function of environmental factors. It therefore becomes possible to test, on ancient populations, models that attempt to predict, for example, the influence of climate change on populations.

Other fields have also been successfully tackled. For instance, the history of human diseases was studied with paleoparasitology approaches. By amplifying sequences of parasites in humans, the diet of past populations has been studied with archaeo-traceability by determining the vegetable and animal species they ate on the basis of crusts in amphora or jars.

Last but not least, paleogenomics enables the reconstruction of the evolutionary history of plants, both throughout their 150 million years of evolution and during their more recent domestication. Comparing contemporaneous genomes with their common ancestors has enabled the identification of molecular mechanisms that gave birth to modern plants as well as the biological functions of the vegetable realm thanks, among others, to the recurrent occurrence of gene duplication [ZHA 14]. Beyond the evolutionary aspect of these results, major economic benefits are expected, as we will be able to improve agronomic features or resilience to environmental stress.

Although paleogenomics might initially have seemed to study anecdotal objects, 30 years later this discipline has gained momentum and become notorious thanks to recent methodological and technical developments such as NGS technologies. Nowadays, paleogenomics contributes to the resolution of a great number of issues in environmental sciences and in the study of biodiversity and its origins.

9

Functional Ecology
and Population Genomics

The association of functional ecology and population genomics enables us to relate the knowledge we have of traits linked to essential functions of organisms to the novel knowledge we are acquiring about the structure and expression of genomes. The information collected by the various omics approaches (genomics*, transcriptomics*, proteomics*, metabolomics*, etc.) directly sheds light on the functional traits of populations, in tight relation with their interaction with the environment in its biotic and abiotic dimensions. NGS technologies are at the basis of the emergence of new analysis approaches called "ecogenomics*" and "reverse ecology*". Beyond functional ecology, these fields of research are expected to propel advances in evolutionary biology, taxonomy, community ecology and chemical ecology, as well as in genomics, physiology, integrative biology, ecotoxicology and ecological engineering.

Knowing the functional properties (or functional traits) of organisms (eukaryotes, bacteria, archaea, etc.) and their viruses is a major stake of functional ecology. This knowledge of functional traits is part of the understanding of the interactions of organisms with their environment in its biological, chemo-physical, spatial and temporal dimensions. These traits must be studied in an integrative approach for it involves their molecular determinants (genes, m-RNA, proteins), their temporal and expressed variability as well as their roles in the life cycle and interactions of organisms considered at the individual or population level.

9.1. NGS technologies and the new approaches of ecogenomics and reverse ecology

The most obvious benefit of NGS it that it allows access to the genomes of a great variety or organisms. Before the emergence of NGS technologies, the genomes of only a few organisms were available: they were mostly organisms that biology and genetics considered as models, such as the human, the mouse, the thale cress *Arabidopsis thaliana,* the nematode *Caenorhabditis elegans* or the bacterium *Escherichia coli.* NGS technologies opened access to the genomes of other organisms that can now be studied by ecologists for their functional traits. On the interface between functional ecology and genomics, the field of genomic ecology or ecogenomics was developed to "study the functioning of the genome in order to understand the relations between an organism and its biotic and abiotic environment" [VAN 12].

Another strength of NGS technologies is that they provide access to the genomic and transcriptomic data of several individuals enabling the integration of the population dimension into the studying of functional traits. This approach allows the study of the variability of functional traits within a population, so this field of research is called "population genomics". The reverse ecology [LI 08] approach was, thus, born from the use of NGS data coming from population genomic studies. Reverse ecology analyzes the variability and conservation of functional traits within populations. The term was built as an analogy to reverse genetics, an approach that compares individuals with and without a given gene to determine the function that the gene encodes. Reverse ecology compares the genomes of individuals of at least two different populations to identify their distinctive genetic traits and, therefore, predict the distinctive functional traits of the populations. This paradigm can be extended to the comparison of transcriptomes. This approach should be able to predict traits specific to a population or a species and is, therefore, of great interest for ecologists, taxonomists and evolutionary researchers.

The knowledge acquired about the molecular determinants underlying known or predicted functional traits must be situated within the networks of genetic and epigenetic regulation of individuals and populations, which are subject to biotic and abiotic environmental constraints. The predicted traits must be characterized at the cellular and physiological levels. These investigations are all the more crucial as the genes and proteins associated with predicted traits often do not have any known function in the databases. Such molecular approaches can only be developed if strong interactions between ecology, genomics, biochemistry and biology are fostered, combining the various omics with predictive or experimental approaches. Then, the integrative characterization of functional traits is a stake that involves fields of expertise that extend beyond the functional ecology community *stricto-sensu*.

The ecogenomics approach is illustrated by the following example of the search for symbiotic determinants of the bacteria *Frankia* and of the actinorhizal plants (*Alnus* and *Casuarina*). The symbiosis between bacteria of the *Frankia* genus and the host plants *Alnus* and *Casuarina* accounts for about 15% of biological nitrogen inputs on Earth, but the involved plants and bacterial genes are hardly known. The research strategy combined:

1) the sequencing of bacterial genomes of *Frankia* and their comparison with genomes of phylogenetic neighbours and of bacteria of the *Rhizobium* genus;

2) the production of bacterial and plant transcriptomes during symbiosis compared to when free-living;

3) the analysis of genes and interesting functions by the elaboration of mutants in host plants and in the studied bacteria, following reverse genetics approaches (a mutant in which a gene is inactivated) or gain of fuction genetics approaches (by studying a similar bacteria, *Streptomyces*, that has acquired a gene of *Frankia*).

This research strategy revealed that the bacterial genomes of *Frankia* do not have nodulation genes like those that have been described in the symbiotic bacteria of the *Rhizobium* genus and that form root nodules on legumes (Leguminosae). In parallel, a transcriptomic approach has shown that several genes of *Frankia* were potentially involved in the symbiosis and that they were diversely distributed in the genome. However, the comparative genomics study of the host plants showed that *Alnus* and *Casuarina* both almost always possess genes that are homologous to those involved in the symbiosis of leguminous plants with *Rhizobium* bacteria [HOC 11]. In particular, the study showed that a calcium pulse was emitted as a response to the addition of the *Frankia* bacteria [GRA 15], a pulse whose function is to trigger the transcription of genes linked to the symbiosis. Furthermore, the transcriptomic approach has shown that all the genes of host plants were over-expressed during the initial stages of the symbiosis [HOC 11]. Among all these genes, there are genes homologous to anti-microbial peptides (called defensins) that are potentially involved in the development of the symbiotic bacteria *Frankia*. One of these defensins is linked to the vesicles, the cells of *Frankia* that are in charge of nitrogen fixation. The porosity of the vesicle membranes or walls modified by this defensin, thus, allowing the glutamate and glutamine amino-acids to exit [CAR 15]. This is the first proposed model of the *Frankia-Alnus* interaction.

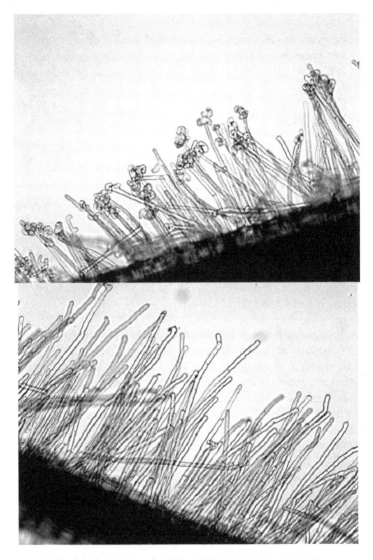

Figure 9.1. *Top: root hair of alder trees treated with the culture supernatant of* Frankia alni, *showing the many deformations that occur during the first stages of symbiosis. Bottom: root hair of untreated alder trees,* © Petar Pujic

9.2. NGS-based analysis of functional traits: reconciling functional ecology and taxonomy

The analysis of the functional traits of populations within complex networks of ecosystemic biological interactions is another major challenge that ecogenomics can solve. Population ecology must be able to apprehend the complexity of ecosystems in which the populations whose traits are studied live. This type of issue stimulates intense interaction between population ecology and community ecology researchers, especially when organisms and functional traits are embedded in trophic networks, geochemical cycles, sustained relationships between organisms (mutualism, parasitism, host or storage transmission) or even when organisms are studied for their effective traits that are estimated to be determinant for the functioning of an ecosystem. Thus, the positioning of populations within complex biological systems is also a vector of expertise aggregation.

Primary data (genomes) and secondary data (identification and characterization of functional traits) that come from ecogenomics are of interest for some fields of evolutionary biology, taxonomy and biodiversity studies directly. For example, evolutionary biology and functional ecology communities can be interested in the same molecular determinants, studying them either for their function or for their adaptive value, and therefore set up a study on the same population. A reverse ecology approach can also enable the identification of characteristic functional traits of a taxon and therefore contribute to its definition. Finally, the characterization of traits and populations contributes directly to decrypting biodiversity.

In the example below, a reverse ecology approach, combined with reverse genetics, enables the research and discovery of the ecological niches of the bacterial species, niches that lead to their speciation.

Following the hypotheses of [COH 06] on the effect of selection on the homogeneity of bacterial populations, we assume that the bacterial species that are defined on the basis of their genomic proximity [STA 02] are actually ecological species that are adapted to specific ecological niches that allow the host species to be adapted to particular dimensions of the environment. Discovering these potential specific niches enables us to understand the speciation mechanisms while pointing to primary ecological niches that enable the bacterial species to evade interspecies competition. This example consists of testing this hypothesis by discovering and characterizing the species-specific genes of *Agrobacterium spp.* This genus consists of numerous species that, although genetically clearly different, are closely related. Furthermore, *Agrobacterium* species used to live sympatrically forming pluri-specific assemblages in the same biotopes, and thus must be able to evade the competition between them.

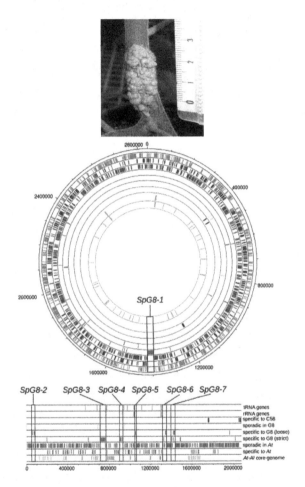

Figure 9.2. *Top: tumor caused by* Agrobacterium fabrum *on a tomato stem,* © *Denis Faure. Bottom: specific regions (SpG8-1.... SpG8-7) of the species* A. fabrum, *as found on the circular chromosome and on the linear chromosome,* © *Xavier Nesme. For a color version of this figure, see www.iste.co.uk/faurejoly/genomics.zip*

The first step (comparative genomics) was to identify the specific genes of the species *Agrobacterium fabrum* by comparing the genomes of several individuals belonging to several related species. The second stage (reverse genetics) consisted of experimentally testing the functional traits predicted by the DNA sequence of the specific genes of the species. This work linked potential specific ecological niches to the species *A. fabrum* [LAS 11].

Comparative genetics analyses showed that there were 196 genes specific to the species *A. fabrum*, since they are absent from other species of the genus *Agrobacterium*. These genes are grouped into seven genomic regions located on the circular and linear chromosomes. These regions code for specific metabolic pathways from which it was possible to infer a hypothetical potential specific niche. This niche would be in the rhizosphere of plants where *A. fabrum* could use plant compounds such as phenolics and specific sugars, overcome iron deficiency, resist toxic compounds and perceive the signals emitted by the plant. The functional genetics studies confirm these predictions and show that the specific genes of *A. fabrum* actually enable its members to evade competition with related species [CAM 14].

This approach, which combines both functional and comparative genomics with reverse genetics, has shown that the bacterial species as defined by genomic characteristics truly is an ecological species and that its speciation, at least in *Agrobacterium*, is parapatric* in type. This approach also reveals the nature of primary ecological niches – most often misunderstood – as niches that allow bacterial species to evade interspecific competition [LAS 15]. Generalizing this approach to the various bacterial species to take into account their ecological role in microbiomes seems useful and productive. Ultimately, it seems essential that the international committee on systematics of prokaryotes (ICSP) requests the genome sequencing of newly described species and encourages this work to be carried out for the more ancient species. This would imply ascribing ecological annotations to each bacterial species, which would be very useful to explore the functional potential of microbial community through metagenome analyses.

9.3. NGS analysis of functional traits as biomarkers of environmental change

NGS technologies bring about novel knowledge and tools to serve societal needs and issues addressed by functional ecology. Applied developments can consist of:

1) estimating or predicting the impact of climate change or of anthropic activities on populations and their functional activity;

2) proposing assessment tools (bioindicators based on populations or biological activity) of these environmental or anthropic changes, with, among others, methods pertaining to the ecotoxicology;

3) using the knowledge about the functional traits of populations to develop ecological engineering or agro-ecology approaches.

The example below shows how NGS technologies can be used to understand the impact of environmental change on the lives of organisms, here the annelids worms.

Failure of the immune system, worsened by environmental change and an increasing number of exotic pathogenic strains that are more or less resistant to many antibiotics, is one of the most serious threats of extinction of marine species. This example illustrates the impact of the environment on the immune system of simple marine organisms (small size and little anatomical complexity) that are resistant to pollution and for which it is possible to work on a significant number of individuals: annelid worms. The worms of marine areas have undergone significant adaptation as they live under extreme environmental constraints, be they thermal or chemical, found in estuaries and in the deep hydrothermal habitats of ocean ridges. The high variability of these environments over short spatial and temporal scales enables to understand with simple, natural-environment experiments, the physiological and genetic responses and the organism-environment interactions.

Capitella, a ubiquitously distributed coastal annelid, is used as a model in ecotoxicology because of its strong resistance to pollution. Interestingly, it was observed that some *Capitella* worms collected from areas showed a microbial growth at the integumentary level, whereas no microorganism has been found on the bodies of *Capitella* that live in a clean environment. This microbial colonization in a eutrophized environment expresses either the acquisition of an epibiose* (beneficial effect) or a microbial infection state (deleterious effect). In both cases, this implies a modification of the immune system of the host: it has either become tolerant to the establishment of a microbial symbiosis or become sensitive to the development of a pathogen it is not able to eliminate. A transcriptomic RNA-seq analysis enables the identification and quantification of genes that are expressed differently in different conditions:

1) genes of populations of *Capitella* that are exposed or not exposed to pollutants;

2) genes of populations of *Capitella* that are exposed to pollutants and that are colonized by microorganisms or not. Through comparative analysis, the study consists in obtaining information about the *in situ* impact of pollution on the antimicrobial immunity of marine organisms.

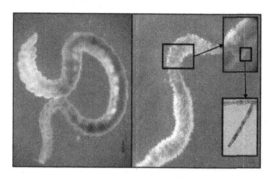

Figure 9.3. *Part of the population of* Capitella *inhabiting anthropized habitats are associated with giant sulfo-oxidizing bacteria (right). No association is observed in non-polluted area (left), © Aurélie Tasiemski*

9.4. Conclusion and perspectives

In the medium term, the evolution of NGS-based functional ecology can now develop along three major paths:

– the first follows a population approach. The number and diversity of organisms whose genomic and/or transcriptomic data becomes accessible will multiply; the data produced will no longer be limited to single individuals but will include several individuals in each population of interest; the study of functional traits will be based on the study of both model organisms and model populations;

– the second focuses on the interactions between the individual and its environment. Functional ecology will be challenged to develop integrative approaches to the knowledge of the traits of organisms with regards to their interaction with the biotic and abiotic environment; the interfacing of functional ecology with biology, evolutionary biology, taxonomy and community ecology will intensify, to eventually become indispensable to develop valuable usage of the produced data;

– the third is about human–environment relationships. Functional ecology will be called upon to address societal issues: ecotoxicology, ecological engineering and restoration, biodiversity, species introduction, control of invasive species, agro-ecology, etc.

The main limitations to knowledge development in this field are the following:

– although the technologies for the acquisition and comparison of genome structures are now accessible, those for transcriptomic data are still being developed (RNA preparations, processing of NGS data). In addition, the acquisition of transcriptomic data is entirely dependent on the environmental conditions in which the studied populations live. Therefore, as the complexity of the environment increases, the difficulty to control (reproducibility) and interpret this data increases;

– the comparing and relating of data from the various omics still face limitations, often for cost reasons or technical constraints such as the development of bio-computing interfaces to the databases, as well as for the lack of incentives to federate the various fields of expertise involved;

– finally, a thorny issue that can become a major challenge in functional ecology is raised by the accumulation, in the databases, of genes and proteins whose functions remain completely unknown. This is a real obstacle to the analysis of functional traits of individuals and populations, but also to the study of metagenomes and metatranscriptomes in the context of community ecology. The decoding of unknown functions is a high-risk activity that is moreover expensive in terms of time and resources. It involves multiple fields of expertise and cannot be undertaken within the framework of short-term projects.

10

Structure and Functioning of Microbial Ecosystems: Metagenomics and Integration of Omics

Associating metagenomic and metabarcoding approaches that describe microbial community structures to functional approaches (metatranscriptomics, metaproteomics, metabolomics, etc.) improves our understanding of how microbial ecosystems work, especially because they enable us to take into account organisms that resist isolation and culture under laboratory conditions. In fact, these culture-recalcitrant organisms may account for most of the microbial community whose functions are essential to the sustainability of our environment and its associated ecosystemic services. Thus, by integrating several organizational levels of specific and functional biodiversity (molecules, genes, populations, communities, etc.), environmental genomics stirs up a synergy between biologists, geochemists and bio-computing scientists, which fosters a better understanding of the relations between biodiversity and functions within a trophic network, in relation with biotic and abiotic factors of the environment [BER 08, SIM 11].

Due to the high-throughput of the new genomic technologies, their use in the environmental field can significantly expand the scope of research in comparison with the classical techniques of population and organism biology. Since they are supposed to be easily applicable to all types of biosphere ecosystems, taking into account all the biological components including viruses, these technologies should enable the tackling of issues involving biological (from the gene to the community) and ecosystemic (spatio-temporal variations) changes as well as the production of data that integrates even the shortest generation times. Considering the various scales and finding appropriate interfaces for them are the fundamental challenges of environmental microbiology.

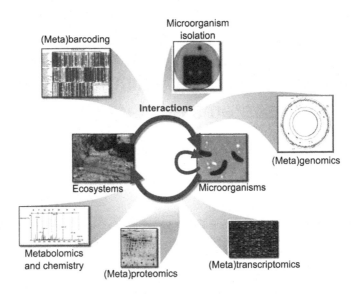

Figure 10.1. *The various aspects of genomics applied to microorganisms and ecosystems. These combined approaches enable the production of an integrated image of their structures and functioning, © Frédéric Plewniak*

10.1. Structure of microbial communities

In addition to technologies that can produce genetic footprints (metabarcoding), the depth sequencing allowed by the new technologies of genomics allow the taking into account of species whose proportion in complex communities is usually underestimated by counting approaches that require the cultivation of organisms. This depth sequencing also provides access to the least abundant species of an ecosystem: the rare biosphere. In microbial communities, these rare species are a reserve of new genes whose evolutionary and functional potential can ensure the sustained structure and functioning of the ecosystem and play the role of a real ecological insurance guaranteeing the resilience, stability and sustainability of our environment.

The following examples illustrate the use of NGS technologies to analyze microbial communities (Example 10.1), the notion of rare species in an ecosystem (Example 10.2) and the biogeography of marine microorganisms (Example 3).

EXAMPLE 10.1.– Analysis of community structure with metabarcoding of the *rrs* (18S rRNA) gene.

To understand the functioning of an ecosystem, a thorough knowledge of the species it hosts and their metabolic potential is essential. For a long time, diversity studies were performed by microscopic identification or by cultivating species. This cultivating approach can be long, tedious, expensive and, above all, only enables the identification of the tip of the iceberg of biodiversity. It was only in 1980 that molecular biology, using first-generation sequencing coupled with phylogenetic analyses, enabled the identification of species without having to cultivate them first.

Recently, diversity studies have benefited from the development of NGS technologies, as they provide direct access to the structural and/or functional diversity of microbial communities. The structure of eukaryotic microbial communities has, thus, been analyzed in the aquatic marine environment by microscope observation (morphological data about organisms) and by molecular approaches (metabarcoding). The complete genomic DNA was extracted from marine samples, amplified by PCR with universal eukaryote primers targeting hyper-varying regions of the gene coding for 18s rRNA. This metabarcoding method, coupled with the analysis of powerful bio-computing technologies, enabled the exhaustive identification of the composition of communities of eukaryote microorganisms [JOB 12, MON 12]. This molecular diversity was then compared to the morphological data collected with the microscope. Although the molecular approach captured relative frequency variations of the most abundant taxons that were consistent with the microscope findings, the two approaches are complementary when estimating diversity because some species were only identified by one or the other method. Microscope observations can identify species whose rRNA is not sequenced or poorly amplified by the couple of chosen primers. Conversely, the molecular method revealed species for which there was no morphological description or for which the morphological description was not sufficiently taxonomically discriminating. Rare species or species that require specific preparations to be visible with the microscope could only be detected with the molecular approach.

Figure 10.2. *Sampling with a plankton net, © Wilfried Thomas/CNRS Photothèque. 20130001_1639*

The coupled use of classical methods and recent molecular techniques increases the accuracy of estimations of ecosystems' structural diversity. These approaches, however, do not provide any knowledge about the role of microorganisms in the community. A metagenomic approach is essential to highlight the diversity of functional genes of a community as well as their seasonal variability. The challenges then consist of optimizing and comparing the data produced by molecular biology (metagenomics and metabarcoding) and microscopy, to evaluate the composition of communities and relate sequence data to microscopic morphology. This step is crucial to understand the trophic networks and the global functioning of an ecosystem.

EXAMPLE 10.2.– The rare microbial biosphere, Archaea in the marine environment.

The determination of the community structure (abundance, richness and composition) of an ecosystem is a central issue in ecology. In marine samples, NGS was able to detect 10 times more OTUs belonging to Archaea than produced by Sanger sequencing approaches. In addition, the greatest part of biodiversity is represented by OTUs that, although numerous, each occur only in small amounts, thus constituting the rare biosphere of microbial communities. The strategy is based on high-throughput sequencing of amplicons* (454-Roche). To avoid biases in the evaluation of richness, a known sequence of the small rDNA subunit, non-existent in the studied ecosystem, is introduced before amplification, to an amount of about 1% of the whole rDNA. This internal standard enables the detection, after sequencing, of errors due to sequencing and amplification. Phylogenic affiliation can highlight clades that belong to the rare biosphere but were never previously detected.

The analysis of the metabarcodes from gene *rrs* 18S amplicons and from the rRNA of the same gene (after a conversion in cDNA) identified active microorganisms within the analyzed communities. The activity of these microorganisms (studied with rDNA *vs* rRNA) showed three distinct fractions: one fraction is only seasonally active and abundant, thus representing a pool of dormant cells (*seed bank*) that are necessary to the functioning of the ecosystem. The second fraction is always rare and active and the last fraction, inactive, includes sequences that have little similarity with databases as it is in fact made of Archaea from outside the ecosystem. The spatial repartition (i.e. biogeography) of these rare microorganisms depends for the most part on geographic distance rather than on environmental conditions.

Studies on the rare biosphere highlight a community that was unknown until recently, that is partly active in ecosystems and that can have a limited geographic repartition. These organisms can carry a yet unknown gene pool whose dynamics needs to be determined. The rare species could have a functional role in the ecosystem as themselves (active rare) or via their dynamics (transition from rare to dominant). These transitions would then be essential keys to predict the functioning of ecosystems in the context of global change [HUG 13].

EXAMPLE 10.3.– Distribution of *Synechococcus* in the oceans.

The Tara-Oceans campaign, which roams the world's seas to collect water samples at various depths in order to study the sea's morphological and molecular biodiversity, has achieved a first comparative genomics and metagenomics analysis of marine picocyanobacteria. *Prochlorococcus* and *Synechococcus* are the two most abundant photosynthetic organisms of the oceans, and therefore play a key role in the global primary production. One of the main research goals is to better understand the relation between genomic diversity within these genera and the ability of various ecotypes to adapt to specific ecological niches of the marine ecosystem [DUF 08].

In collaboration with the Genoscope in France, 25 *Synechococcus* genomes have recently been sequenced, raising to 57 the number of marine picocyanobacteria genomes currently available (40 *Synechococcus*, 14 *Prochlorococcus* and 3 *Cyanobium*). A comparative analysis of 40 genomes of the genus *Synechococcus* highlighted the presence of genomic islands specializing in adaptation to specific ecological niches, including 4 to 6 gene islands involved in acclimatization to different light qualities.

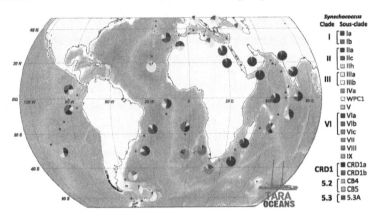

Figure 10.3. *Analysis of the distribution of phylogenetic groups of marine* Synechococcus *at 32 stations along the transect of the oceanographic Tara-Oceans cruise. The assignment of a particular sequence to a taxonomic group (sub-cluster, clade or sub-clade) was done by recruiting environmental sequences of the 40 reference genomes of* Synechococcus, © *Laurence Garczarek. For a color version of this figure, see www.iste.co.uk/faurejoly/genomics.zip*

These genomes have been used as references for analysis of the metagenomic data provided by the TARA-Oceans campaign, by recruiting sequences of natural populations of cyanobacteria on the nearest genome, thus enabling their functional and taxonomic assignment [PIT 14]. These analyses have highlighted:

1) the predominance in some ecosystems of clades that had hitherto been considered as minorities;

2) sudden changes in community composition between different ocean basins, particularly between the Mediterranean and the Red Sea or on either side of the Cape of Good Hope. The availability of such a set of reference genomes also highlighted that not only populations belonging to the same clade could be distributed differently, but also populations of different

sub-clades. Thus, this approach opens access to a much more finely resolved taxonomic scale than provided by the gene 16S RNA, and should enable researchers to better define the notion of ecotype (or species) within marine picocyanobacteria.

10.2. Revealing the functioning of microbial communities

The analysis of the functioning of microbial ecosystems is facing two major shortcomings in the overall context of environmental science:

1) the lack of ecological microbial theories that overlap wide-ranging ecological theories; 2) the lack of biogeographic and temporal patterns that would intersect – or not – those found in some animals and plants. These shortcomings have considerable consequences. For example, they result in the fact that microorganisms, above all viruses, are not really taken into account in addressing global issues (CO_2 and other greenhouse gases, climate, acidification, ozone, eutrophication, hypoxia, invasion, xenobiotic dispersion, etc.) because of our ignorance of microbial patterns at the time, space and organizational scales. Within the overall framework of organism biogeography, the relating of molecular sequences with microscopic phenotypes is a major technological and scientific challenge. NGS approaches help to reveal the key role of microorganisms in the dynamics of ecosystems, as the following examples highlight.

10.2.1. Metatranscriptomics reveals the functions of eukaryotic microorganisms

The metatranscriptomics approach is the preferred approach to the study of eukaryotic microbial communities such as protists, fungi, and mesofauna (nematodes, acari, springtails, etc.). These organisms play a key role in the functioning of aquatic (photosynthesis, primary consumers) as well as terrestrial (vegetable biomass degradation, pathogens and symbionts of animals and plants) environments. The great size and complexity of eukaryote genomes make it difficult to annotate environmental DNA sequences of eukaryotic origin, for they are characterized by many introns and the abundance of repeated sequences, including transposons. Eukaryote mRNAs, which exhibit the unique property of being polyadenylated (i.e. mRNA having additional A bases at the end) can nevertheless be specifically isolated and analyzed separately from environmental RNA (a majority of ribosomic RNA and bacterial mRNA).

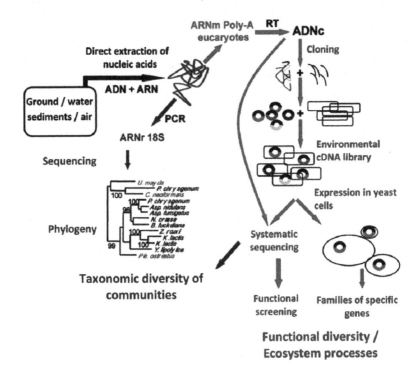

Figure 10.4. *The various aspects of the metatranscriptomics approach. mRNAs and rRNAs of eukaryote origin are extracted from an environmental sample; mRNAs are treated with reverse transcriptase (RT) in order to obtain cDNAs that undergo systematic sequencing or are cloned into expression vectors that enable functional screening in yeast; rRNAs are treated with RT, then cDNAs are sequenced and analyzed to describe the diversity of taxons found in the analyzed sample, © Roland Marmeisse. For a color version of this figure, see www.iste.co.uk/faurejoly/genomics.zip*

High-throughput systematic sequencing of cDNAs is the preferred approach to apprehend the complexity of metatranscriptomes. Current technology generates short sequences (<500 pb, and even only 100-150 pb), and therefore assembling them into complete genetic sequences is challenging. This is possible for microbial communities that are dominated by very few taxons (for instance in very polluted environments) or for microbial communities where a significant scientific community engages in the sequencing of many reference genomes (for instance, the human microbiome). Although the sequencing of metatranscriptomes now leads to editing lists of genes to which functions can be associated, associating these genes to a taxonomic origin still seems far-fetched. Beyond the improvement of bio-computing analysis performance, it is important to support initiatives that aim to systematically sequence the genomes of species representing the diversity of life, in order to associate observed functions with taxonomic groups.

Which level of reliability can we grant automatic *in silico* annotations, and what about the numerous sequences without homologous counterparts in databases? In addition to systematic sequencing, it is necessary to promote the experimental analysis of environmental sequences likely to reveal new biological processes within ecosystems. For example, by expressing the cDNAs of eukaryotes living in soil in yeast strains, a new family of peptide transporters that fungi may be using to collect organic nitrogen from soil was described [DAM 11], whereas the *in silico* analysis of these sequences predicted that they were amino-acid transporters. The same approach also characterized new genes for resistance to heavy metals, which is important in microbial communities that have to adapt to the pollution of the environment [LEH 13].

To better account for the complexity of the biological processes which unfold in ecosystems, metatranscriptomics and metagenomics must be combined with complementary functional ecology approaches such as measures of nutrient flows and storage.

10.2.2. From multi-omics data to the metabolic model of the community

This example shows how complementary experimental and modeling methods can be combined to study the trophic interactions of a eukaryotic (*Euglena mutabilis*) microbial community isolated from the Carnoulès (Gard) site of acid mine drainage and of a bacterium ("*Candidatus* Fodinabacter communificans") that was found on the same site. "*Ca.* Fodinabacter communificans" is one of the seven dominant strains identified at the Carnoulès site and whose genome was reconstituted by metagenomic sequence assembly [BER 11]. Proteomic and metabolomic approaches carried out on *Euglena mutabilis* have shown that this protist secretes into the environment metabolites that may be used by other microorganisms [HAL 12], in particular, by "*Ca.* Fodinabacter communificans". This bacterium, and possibly others, may in turn provide it with the cobalamine it is not able to synthesize itself.

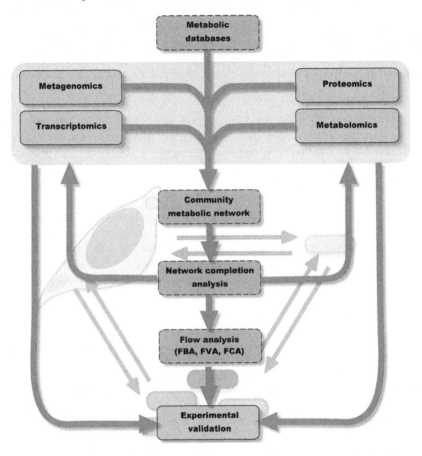

Figure 10.5. *Modeling strategy within a community on the basis of multi-omics data. Frames with solid outlines represent experimental aspects (data production and validation), frames with dashed outlines correspond to bio-computing data and analysis, © Frédéric Plewniak*

A model of the combined metabolic network of the two organisms and of their environment is being devised. It consists of modeling, for the first time at the metabolic level, on the basis of functional genomics data, the interactions in the environment of organisms whose genome is not entirely or not at all sequenced. The role played by additional players in the exchanges will also be taken into account. The model relies on the integration of metagenomic, proteomic, metabolomic and transcriptomic data. Computationally, graphical analyses are used to identify the important and coupled reactions. The study will consist of characterizing the interactions between *E. mutabilis*, "*Ca.* Fodinabacter communificans" and their environment beyond individual mechanisms of resistance, in addition to highlighting the exchanged metabolites if present, and in documenting the potential connections between their metabolisms.

10.3. Conclusions and perspectives

Coupled with the necessary laboratory and *in situ* experimentation, NGS technologies applied to microbial ecosystems also require the upstream development, of sampling and extraction methods, and downstream of databases and methods enabling not only the storage and exchange of data in standardized formats but also – and above all – its integrated analysis (assembly, annotation, comparison, etc.), especially in the case of the short-length reads that are produced when the environment is complex.

Environmental genomics applied to the functioning of ecosystems should thus enable, beyond the mere listing of biological inventories (species, genes, etc.):

1) to take into account the characteristic symbiotic associations of organisms in an environment (interactomics*, especially in the molecular dialogue of prokaryote-eukaryote systems);

2) to apprehend the adaptive processes and the potential role of the genomic plasticity of microorganisms;

3) to link the dynamics of these ecosystems with their associated social, economic and cultural development;

4) to anticipate the evolutions of ecosystems in response to constraints, including anthropic ones, with predictive models developed on the basis of our understanding of their properties.

11

Modeling and Predicting Behaviors and Dynamics of Ecosystems

The functioning of ecosystems is nowadays understood as the result of interactions between various micro- and macroscopic organisms. These interactions, especially between microorganisms, have significant impacts on ecosystemic services (e.g. nutrient recycling, production of organic matter etc.). Although microbial abundance is a long-known fact, high-throughput genomic data have highlighted an amount of diversity that was unsuspected [ROE 07, HIN 13]. These novel descriptions, however, remain one step towards the goal of elucidating, at the molecular level, the functioning of microbial ecosystems. To reach this goal, using dedicated models that mobilize the whole diversity of biotechnology resources is of fundamental importance [ZEN 12].

Metaomics approaches provide access to several information levels that are all useful for modeling. Metagenomics provides access to the diversity of actors in the system that has to be modeled, metatranscriptomics and metaproteomics provide access to the functions expressed in a given community. It is, therefore, now possible to have a complete understanding of the biogeography of the microorganisms of an ecosystem on a whole territory and even from a holistic perspective (i.e. in all its dimensions) [KAR 11].

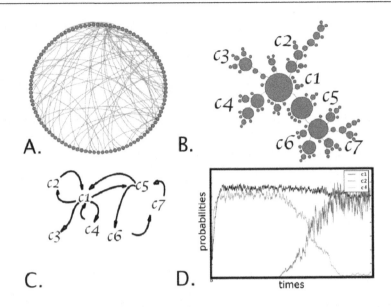

Figure 11.1. *The various steps for modeling on the basis of high-throughput data; a) identification of correlations between communities; b) identification of communities; c) identification of interactions between communities in order to model d) the probability of activation of communities depending on inter-dependencies. Each one of the four steps represents an existing field of research, the cohesion of these four steps is the major challenge facing the modeling of environmental microbial systems,* © *Damien Eveillard*

11.1. Intensive computation benefits the description of biodiversity with metagenomics

Intensive computation is often associated with metagenomic approaches such as metabarcoding for the inventory of microbial communities. The aim is to develop computer science technologies to automatically build inventories of microbial communities based on the sequencing of NGS-generated amplicons. The following example accounts for the study of diatoms, a group of unicellular algae species that makes up most of plankton.

Diatoms possess a silica skeleton (the frustule). These organisms are free living organisms or in biofilms, either in marine or freshwater environments. They can reproduce either sexually or by cloning. The communities of diatoms (there are about 20,000 species) are used as bio-indicators of the quality of the environment. Diatoms are an excellent biological model because a diversity inventory, which is a prerequisite to any microbiological study, can be realized either in a classic naturalist manner (based on morphological identification) or by molecular approaches such as barcoding. Inventories based on morphological identification of frustules are very tedious. It is nowadays possible to use metabarcoding to identify and quantify these microorganisms. This approach, however, requires researchers to overcome algorithmic issues related to time and space limits memory and time limitations.

Figure 11.2. *Frustules of* Nitzschia sinuata *observed using a scanning electron microscope. This common species can be found in thermal springs of the Massif Central in France. The frustule is made of two overlapping parts, or thecae, that form a sort of box that can expand. The scale is given by the white bar on the side, which represents 10 micrometers, © Geneviève Bricheux/CNRS Photothèque 20080001_0480*

Figure 11.3. *Biofilm on muddy substrate, formed
within hours during ebb tide,* © *Dominique Joly*

The classic NGS-based identification strategy consists of producing reads* of markers of taxonomic interest (rbcL, 18S, cox1 for diatoms) and to classify them using either supervised methods like BLAST, comparing them to a complete reference sequence in a database or unsupervised methods with a set of classification processes [SCH 11]. The reference database is available within the framework of the RSYST network (http://www.rsyst.inra.fr/). A previous study tested the accuracy of the BLAST method for NGS data by analyzing the results obtained from pyrosequencing of artificially constructed communities from cultivated strains, whose composition was well known. Although BLAST allow to identify existing strains, it also produces false positives (Type 1 errors) by identifying strains that were not present in the communities. Thus, exact alignment algorithms were developed on the basis of global (Needleman and Wunsch) and local (Smith and Waterman) alignment algorithms. Exact algorithms are more time and memory consuming but provide more accurate results. Although global alignment can be achieved in a reasonable time on a desktop computer, computing all local alignments requires massive computing capacities. Since it is a massively distributed problem, grid* computing is an adequate technology for this task, although it requires significant learning and training investments for users. So, this field of research has to progress along three paths:

1) developing supervised classification algorithm and taxonomic inventory technologies that rely on local exhaustive alignments;

2) developing parallel unsupervised classification technologies;

3) facilitating access to grid computing for the community of biologists. This general strategy, which can be applied to other similar approaches, requires a tight collaboration between biologists, applied math scientists and computer scientists around intensive computing technologies.

11.2. An emergent scientific domain: the ecology of systems

Merely describing ecosystems is sterile if no effort is made to integrate various heterogeneous pieces of data as provided by research communities, using a dedicated modeling framework allowing to study the ecosystem as a whole. These complementary approaches paved the way to a novel discipline of ecosystems modeling called "systems ecology" [KLI 11]. Systems ecology improves the description of biological compartments within ecosystems, in terms of diversity as well as functionality and provides access to higher levels of complexity. Relying on technologies developed for the "systems biology" domain (that deals with lower levels of organizations, such as cells, organs or organisms), the role of computer science is fundamental and goes much beyond issues of computing power and storage capacity as it has the potential capacity to extract emergent properties.

Thus, although metaomics data can initially be used as slightly complex refinements to existing models, the generalization of novel genomic models will rely on the structuring of systems ecology into a fully qualified scientific domain. Like systems biology, it will be based on computer science, mathematics and ecology. This structuring will essentially rely on reference ecosystems that will be usable as testing benchmarks for new techniques.

Systems ecology is based on two complementary approaches:

– a bottom-up approach whose aim is to provide a formal description of metagenomic data (for example, the highlighting of a meta-metabolism of the ecosystem – which must converge to existing biogeochemical cycles);

– a top-down approach whose aim is to enable focused searches in genomic databases to validate existing models and reason on the functioning of ecosystems. This approach will, for example, allow us to isolate species whose genomes encode reactions that cause the main geochemical processes.

The bottom-up approach we present here was implemented to elaborate a descriptive model of biogeochemical processes within the polluted marine sediments of two sites in the Mediterranean Sea [PLE 13]. The goal was to highlight, using metagenomics, biotic factors that could influence the arsenic cycle in sediments. Such an analytical approach is applicable to any issue about the severe pollution of industrial and post-industrial sites, a major risk in terms of public health.

a)   b)

Figure 11.4. *The studied sites of Saint Mandrier*
a) and l'Estaque b), South of France © Wikimedia commons

The port of l'Estaque, near Marseille is a former metallurgy site and is now polluted by arsenic and heavy metals; the port of Saint-Mandrier, near Toulon, has high metal pollution but a much lower arsenic concentration than l'Estaque (Figure 11.4). Metagenomes of sediments in these two sites have been sequenced by 454 Roche GS FLX Titanium. Sequences of these metagenomes have been annotated using RAMMCAP system, available on the portal CAMERA [SUN 11], along with four other control metagenomes. These controls were used to highlight the processes shared by the two polluted sites, because a mere direct comparison would have hidden them. Coupling annotations from Gene Ontology of the two metagenomes to informations about sites geochemical parameters [MAM 13] allowed to develop a model describing the behavior of the two communities, without the need for genome assembly.

Figure 11.5. *Marine sediment sampling in the port of Estaque,*
© *Sandrine Koechler/GMGM/CNRS Photothèque 20130001_1630*

The results show that in l'Estaque the sulfur cycle is central to the model and is coupled to fermentation and the arsenic cycle (Figure 11.6).

Fermentation provides the sulfate-reducing bacteria with the electron donors necessary to sulfate reduction. The produced sulfurs seem to react with arsenic, forming thioarsenized compounds that have been experimentally observed in l'Estaque. These compounds being very soluble, they favor, on this site, the dispersion of arsenic towards the column of water. These compounds being very soluble, they favor, on this site, the dispersion of arsenic towards the column of water.

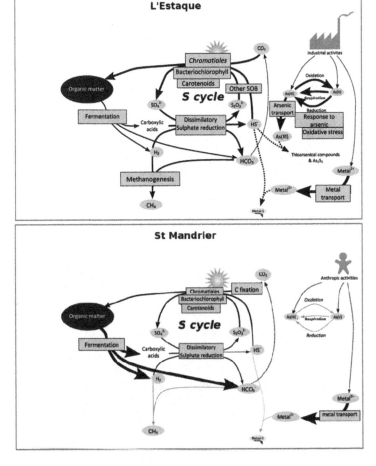

Figure 11.6. *Comparison of the biogeochemical cycles by metagenomic analysis of two Mediterranean sites polluted by heavy metals; (top) the former metallurgy site of l'Estaque, and (bottom) the port of Saint-Mandrier, © Frédéric Plewniak*

This project shows, that it is possible to study complex metagenomes even with a limited coverage and without genome assembly, provided the relevant functional level (biological processes) is chosen. In that case the metagenome is considered as a supra-organism, the global metabolism of which we study. Since direct comparison only provides an incomplete picture, adequate controls have to be used to compare similar metagenomes. Ideally, controls would only involve ubiquitous processes, in order to avoid hiding other important processes. On that account, some standards (i.e. sort of average values for different environmental categories) would allow us to focus on the relevant data, but are lacking, which would enable us to focus on the relevant data. It would also be interesting to go beyond descriptive models to elaborate more complete models that would provide access to more detailed analyses of the relations between various element cycles and of the impacts of the various processes on the dispersion of arsenic.

As illustrated by this, models arising from systems ecology are very complex as they describe the diversity of communities as well as matter and energy transfers within. This modeling approach, when applied to ecosystems of societal interest, will propel the emergence of "synthetic ecology" whose goal will be to define the optimal conditions in which communities can be led to help the natural denitrification of soils, the production of biofuels on basis of controlled ecosystems and so on.

Although metaomics* and functional data is now accessible (for example, via flow measures) for similar spatial scales, these scales remains largely greater to the microbial interaction scales, that are often as precise as the molecular level. Relating diversity and functional data is also difficult. In addition, existing databases are still heterogeneous, hardly standardized and their access by most of the modeling community remains difficult.

11.3. Modeling and predicting

The elaboration of predictive models is still conditioned to the integration of the spatial and temporal dimensions of the diversity and behaviors of microbial communities. Although the integration of metagenomic data allows the construction of causality graphs that describe interactions between microbial communities, it does not give a functional dimension, at the molecular level, to the studied ecosystems. To overcome this issue, modeling techniques need to be developed that integrate other types of quantitative (or functional) information coming from physico-chemical experiments, even if this quantitative information remains incomplete compared to the covering of the metagenomics knowledge.

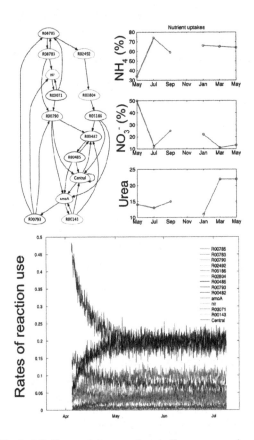

Figure 11.7. *Probabilistic modeling protocol. From a graph of events, (bio-geo-chemical cycle of nitrogen), via probabilistic inference from experimental data (red), the dynamics of the modeled system can be simulated. Here, the variation with time of reaction used to reproduce the experimental behavior,* © Damien Eveillard

To develop such models, a probabilistic approach can be used by applying probabilities to interactions between communities of organisms. [BOU 11], for example, have recently shown that by introducing probabilities in an otherwise boolean model, one could accurately integrate the qualitative and quantitative aspects of a living system. ETG (*Event Transition Graph*) modeling is applied here to a small microbial network that represents the biogeochemical cycle of nitrogen, to quantify the influence of a microbial community on the global function of the ecosystem.

The model considers:

1) the description of a biological event and of its interactions (here the occurrence of a biogeochemical reaction producing a compound, linked to other relations that use the compound as a substrate);

2) a quantitative information such as the variation of a compound with time. Through an optimization process, ETG modeling consists of identifying a set of probabilities to associate to interactions of events, so that the graph, weighted on this basis, reproduces, on average, the quantitative behavior proposed in step 2. Results show that the biogeochemical cycle of nitrogen is mainly controlled by bacterial communities. The corresponding network consists of the successive bio-geochemical reactions triggered by various microorganisms. This modeling approach considers the ecosystem as a meta-metabolic system that synthesizes bacterial communities through the reactions they control. The qualitative graph is made of 14 events or reactions and 32 interactions (Figure 11.6). Each event induces a quantitative effect: each event when taken during the random cruise within the graph; induces an effect of increasing or decreasing the population of the associated population (i.e. an increase of the ammonia-oxydizing population when the *amo* reaction is triggered) and otherwise a passive degradation. To estimate the probabilities of the ETG, the variations of quantities of ammonia and nitrates in the Chesapeake bay (USA) are used [BOU 11]. After machine learning processes, the probabilities enable the model to reproduce variations of ammonia and nitrates, but also to reproduce variations of nitrites even so this concentration was not considered so far. Once the probabilities are set, a specific algorithm simulates the probabilistic model (Figure 11.7) and researchers can reflect on the system in order to identify the relative importance of an event with regards to the others and to "monitor" the quantitative behavior of the microbial ecosystem.

ETG modeling focuses on the quantitative dynamic behavior. This aspect of modeling is supported by the fact that quantitative measures are rarely collected at equilibrium and that functional measures are mainly focused on communities that are under strong adaptive pressures. An extension of ETG is currently developed to take into account diversity as a quantitative measure of learning.

11.4. Conclusion and perspectives

To meet the challenge of integrating spatial and temporal dimensions as well as the diversity of microorganism communities and their behaviors, we need to strengthen metagenomic sampling strategies and use them at the micro-environment scale. Knowledge of the spatial [RAY 14] and the temporal structure of the microbial ecosystem remains scarce compared to the mass of available metaomics* data. In addition, current environmental genomics analyses focus on the correlation of abundance of communities of separate ecosystems, whereas the functional aspects of ecosystems can only be reached via causality information. Lastly, another challenge consists of improving the data sharing strategies to ease the task of making them available on centralized repositories.

Reinforcing the relations between microbiologists, modelers and bio-computing scientists is necessary. Interdisciplinary research is the only strategy allowing us to model the complexity of current issues. This implies reinforcing academic training by, for example, proposing specialized elective courses within the current computational biology and modeling curricula. It also implies breaking down barriers between laboratories and supporting initiatives that foster collaboration between various research units.

To promote interdisciplinary research and the standardization of modeling techniques, it would be useful to identify reference sites and ecosystems (for example, LTER – long term ecological research network). This way, all available (genomic, functional, spatial and temporal) data would be freely available to stimulate the design of computer science solutions. In addition, the various sampling strategies need to be reconciled to maximize the consistency of the representations of genomics with functional knowledge and thus let the causality relations between communities emerge. The benefits of such an integrating effort will include the arising of a strong international enthusiasm for some well-focused issues. Such initiatives are already being carried out on several sites [DAV 12] but they remain too scarce and incomplete. With open data, the various bioinformatics platforms and data centers will be able to federate and foster a necessarily collaborative approach (especially via the setting up of Galaxy-type workflows) to achieve the analysis of the environmental genomics "datanami".

12

The Omics of the Future

The diversity of scientific issues and methodologies described in the previous chapters clearly shows the dynamism of the international research community working on environmental genomics as well as its commitment to propel the environmental sciences into the integrative field of systems ecology. In this respect, omic technologies not only foster renewed scientific practices and questioning but also pave the way for a novel perspective on social and societal issues. The diversity of expertise in this domain calls for promoting initiatives to a systematically interdisciplinary approach of objects and issues dealing with environmental genomics. It also advocates to foster better sharing of data collection and valorization.

This final chapter summarizes the exceptional potential and opportunities of environmental genomics, especially in the challenging fields related to the ecosystem dynamics and functioning, changes caused by humans and, more generally, to the management of the impacts of global change on ecosystems. This chapter also hopes to inspire vocations among the youngest readers.

The future of environmental genomics depends on four complementary developments:

1) The first axis concerns the studied objects. Presently, a significant amount of data on model organisms (some mammals, birds, fish, plants, bacteria or fungi) is available but we are still far from a complete picture of biodiversity, whether living organisms are abundant, rare, remarkable or cryptic. Environmental genomics will increase our knowledge of living

organisms, notably by addressing the poorly-studied taxons for analyzing their diversity, distribution, roles in ecosystems as well as their behavior and interactions with other organisms. These poorly-studied organisms may be macroscopic (like insects, crustacean and fungi) or microscopic (viruses, protozoa, bacteria or helminths), autonomous, parasitic or symbiotic. In this perspective, the range of focused ecosystems is wide, including soils, oceans as well as microbiomes (intestinal, epidermal, foliar, etc.). Significant efforts must be made to coordinate and diversify the acquisition of data about the Tree of Life, not only for taxonomy purpose, but also physiological, environmental and societal knowledge to integrate the interactions and retro-actions between organisms, humans and their environments.

2) The second axis is related to technical challenges as well as scientific and societal issues related to the accuracy of the information provided by omic approaches. Omic technologies generate gigantic data flows which pose the great risk of erroneous scientific and societal interpretations, which poses the great risk of erroneous scientific and societal interpretations. Improving the accuracy of the information is urgent and necessary, at the level of omic data production technologies (technical and methodological, especially for the analysis of complex communities) as well as at the level of the mutual consolidation of the various omic approaches (for example, automatic but expert annotations or the interfacing of data sets). The omics interpretation can also be strengthened by associating other predictive or experimental approaches such as the analysis of trophic relations, ecocomputing or socio-ecology. Lastly, more synergy between the different scientific communities is necessary to understand how the omics "data avalanche" radically changed the effect of human societies on life and human populations themvelves: for examples, the commercialization of technologies that identify the ethnic origin or the effort to quantify of anthropical impacts on biodiversity.

3) The third axis concerns computer infrastructures, including the technologies to analyze, store, interoperate and archive omic data. High-throughput technologies generate massive volumes of heterogeneous data (DNA/RNA, proteins, chemistry of life, geochemistry, chemo-physics) and sources (populations, communities, ecosystems). Beyond issues about the processing, storing and sharing (storing space and computing power) of data, the primary issue of their valorization exploitation is currently only superficially addressed. The data must be available for uses that go beyond

what it was originally produced for, it must be usable by other researchers to address different issues; the data mining from large sets of data raises the issue of the compatibility and interoperability of data and metadata*. Developing compatible methods and defining standards for interoperability is, thus, a challenging question.

4) The fourth axis is about promoting cross-disciplinary approaches and generalizing modeling methods that encourage the data sharing and the cross-interpretation between data sources. To this end, we must identify reference objects, sites or ecosystems for which all the data (either environmental, genomic, functional, spatial or temporal) will be shared to stimulate the development of integrated models.

Figure 12.1. *The main key-words of this book. Statistically, the word "data" is the most frequent of the book. It reflects the current revolution in the environmental sciences, a revolution faced by researchers in the era of "Big Data", in the various domains related to life and technologies, with all the issues and challenges described. As expected, the words "species", "DNA", "sequencing" and "environmental" are, in decreasing order, the next most important words, followed by words that actually describe life, such as "organisms", genomes", "biodiversity", "communities", "populations". Lastly, we see the omic words such as "technologies", "genomics", "methods", intertwined with the previous ones. The "data" word highlights the need for interdisciplinary approaches in environmental genomics and the need of complementary tools of biological and computing skills, © Dominique Joly*

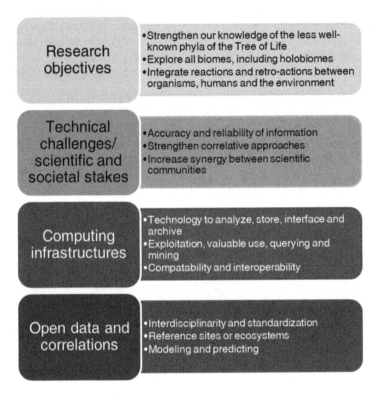

Figure 12.2. *The major challenges facing environmental genomics,* © *Dominique Joly.*

At the institutional level, two important points need to be considered as national and international priorities by institutions and governments. The first point concerns the need for specific training and for the development of conceptual frameworks to engage with the emergent new research issues raised by the omics. The constant growth of interfaces between disciplines, especially for the development of mathematical, bio-computing and statistical techniques needed for the valorization of data will have to promote adaptable and diversified strategies to increase skills of all actors including junior and senior researchers. This should enable the community to keep pace with the progression of technologies, as the threshold of third generation NGS has already been traversed.

The second point is about cumulating the benefits of research infrastructures and programs (for example, via national and international research agencies and with the emerginig countries) that are devoted to environmental genomics. The current priorities for national infrastructures concern domains of economic interest (agribusiness, carbon resources, etc.), public health (cancer, neuro-degenerative diseases, rare diseases, etc.) while environmental genomics receives comparatively little interest. Resorting to private sequencing companies is an insufficient response to this issue. It does not meet the needs for optimization and developments of specific technological procedures that are dedicated to fundamental research. In addition, the financing of research programs remains scattered and limited. Concerted action at the national, international and inter-institution level would certainly be beneficial to the future of this emerging field. In this sense, transnational meta-projects like Future Earth could certainly contribute to foster the international coordination of major program dealing with environmental genomics.

As a whole, these considerations are consistent with the remark expressed in the "Big Data" White paper of [AHA 14], including the promotion of cross-disciplinary coordinated cooperation to promote the setting up of consortia for environmental genomics. This thematic field should be the first advocate of a systems ecology that integrates humans, in their individual, social and societal dimensions, as players that not only act on the environment, but are also subject to its selective pressure. Meeting the challenge that consists of developing a planetary perspective on evolution and life is the real vocation of 21st century research.

Glossary

General Glossary

Alignment: comparison of DNA or protein sequences by placing them one under the other in order to detect their similarities and differences.

Assembly: action that consists of reconstituting a genome on the basis of its sequenced fractions (also named reads). Computer-based assembly is possible due to partial overlaps of reads.

Barcode: sequence of DNA, also called molecular marker, which allows the assignation of an organism to a taxon or a determined group of organisms. The sequence of each barcode is compared to DNA sequences of a reference database.

Coverage: number of times the same DNA region is represented after the assembly of a genome or metagenome. The higher the sequencing coverage is, the more accurate the knowledge of the assembled sequence is, because more sequencing errors have been corrected.

Ecogenomics: research field that uses (meta)genomic and (meta)transcriptomic data to understand ecology and the adaptation mechanisms of organisms.

Epibiosis: biological interaction between two organisms in which one uses the other as a supporting base.

Functional ecology: research field whose object is to study the functioning of ecosystems and organisms.

Genomics/transcriptomics: science that analyzes the genome (DNA) or transcriptome (all the RNA transcripts) of individuals or populations.

Gigabase (and other related units): the total number of A, T, C or G bases that constitute a genome or any DNA fragment as a measure of the size of this genome or DNA fragment. This measure unit is declined in a kilobase (kb) for one thousand bases, a megabase (Mb) for a million bases and a gigabase (Gb) for a billion bases.

Metabolomics: science that studies all the metabolites (sugars, fatty acids, amino-acids, etc.) of a cell, a tissue, an organism or an environment.

Metagenomics/metatranscriptomics: science that analyzes all the genomes (DNA) or transcriptomes (all the RNA transcripts) of a community of organisms.

Metaomics: approach combining omic approaches such as (meta)genomics, (meta)transcriptomics, (meta)proteomics, metabolomics, etc.

Microorganisms: uni- or pluricellular microscopic living organisms that are bacteria, archaea and some eukaryotes, as well as their associated viruses.

Molecular taxonomy (taxon): science that uses genomes or molecular markers to describe living beings and group them into entities called taxons (species, genus, family, etc.).

OTU: Operational Taxonomic Unit. Set of DNA sequences grouped on the basis of their similarity according to an arbitrary threshold. For example, for a 5% threshold, only the sequences that are less than 5% different belong to a same OTU.

Phylogeny: study of the reproductive relatedness of living beings to understand their evolution.

Pipeline analysis: series of bio-computing programs that enable the automatic analysis of the data produced by omic approaches.

Proteomics: science that studies all the proteins of a cell, a tissue or an organism.

Reference genome or reference assembly: sequenced and assembled genome that is used as a reference to assemble the genomes of taxonomically related organisms.

Reverse ecology: approach that consists of predicting the ecology of a population of organisms (and especially its functional traits) on the basis of genome comparisons.

Speciation: mechanism through which populations are differentiated into distinct species. The speciation is sympatric when the repartition areas of the populations are the same, allopatric when they are different and parapatric when they partially overlap.

Technical Glossary

454-pyrosequencing: high throughput second generation sequencing technology.

Amplicon: DNA fragments that result from the amplification of DNA with PCR (Polymerase Chain Reaction).

Contig: computer DNA sequence that is the result of the assembly of DNA reads produced by DNA sequencers.

De novo **sequencing and assembly**: sequencing and assembly of genomes that does not rely on a reference genome, as it only uses reads and contigs produced by DNA sequencers.

Grid computing: software that provides users with almost unlimited computing or storage capacities thanks to a transparent and easy access (a very high throughput network connection, like the Internet) and to a large pool of computing resources that are massively distributed.

Illumina: high throughput second generation sequencing technology.

Interactomics: study of the interactions between various bio-chemical relations. It enables the understanding of protein interaction networks.

Library: large set of physical or software elements, such as clones or gene sequences, grouped in the same location (tube or file).

Multiplexing: technique that consists of analyzing several samples in parallel, for example with one single NGS sequencing operation or one single-PCR amplification.

RAD-seq: Restriction site Associated DNA sequencing. Partial sequencing of genomes on each side of enzymatic restriction sites that enables the comparison of thousands of genetic markers spread on the whole genome. This approach corresponds to partially sequencing the genome; it is used to compare large genomes whose complete sequencing would be impossible.

Read: computer DNA sequence directly coming from DNA sequencers.

Run: complete operating cycle of a DNA sequencer.

Scaffold: assembly of DNA contigs that offer the most compact possible representation of the genome, even if the sequences between the contigs are still undetermined. A genome assembled as the combination of several scaffolds is considered incomplete.

SNP: Single Nucleotide Polymorphism. Variation of only one nucleotide in the compared genomes or sequences.

Bibliography

[AFI 15] AFIAHAYATI, SATO K., SAKAKIBARA Y., "MetaVelvet-SL: an extension of the Velvet assembler to a de novo metagenomic assembler utilizing supervised learning", *DNA Research: An International Journal for Rapid Publication of Reports on Genes and Genomes*, no. 22, pp. 69–77, 2015.

[AHA 14] AHALT S., BIZON C., EVANS J. *et al.*, "Data to discovery: genomes to health. A white paper from the national consortium for data science", *RENCI, University of North Carolina at Chapel Hill Text*, available at http://dxdoiorg/107921/G03X84K4, pp. 1–19, 2014.

[AIM 13] AIME C., LAVAL G., PATIN E. *et al.*, "Human genetic data reveal contrasting demographic patterns between sedentary and nomadic populations that predate the emergence of farming", *Molecular, Biology and Evolution*, no. 30, pp. 2629–2644, 2013.

[ALS 11] ALSHEIKH-ALI A.A., QURESHI W., AL-MALLAH M.H. *et al.*, "Public availability of published research data in high-impact journals", *PLoS One*, no. 6, p. e24357, 2011.

[AND 13] ANDERS S., MCCARTHY D.J., CHEN Y., "Count-based differential expression analysis of RNA sequencing data using R and Bioconductor", *Nature Protocols*, no. 8, pp. 1765–1786, 2013.

[AND 14] ANDREWS K.R., LUIKART G., "Recent novel approaches for population genomics data analysis", *Molecular Ecology*, no. 23, pp. 1661–1667, 2014.

[BAI 08] BAIRD N.A., ETTER P.D., ATWOOD T.S. *et al.*, "Rapid SNP discovery and genetic mapping using sequenced RAD markers", *PLoS One*, no. 3, p. e3376, 2008.

[BER 08] BERTIN P.N., MEDIGUE C., NORMAND P., "Advances in environmental genomics: towards an integrated view of micro-organisms and ecosystems", *Microbiology*, no. 154, pp. 347–359, 2008.

[BER 11] BERTIN P.N., HEINRICH-SALMERON A., PELLETIER E. *et al.*, "Metabolic diversity among main microorganisms inside an arsenic-rich ecosystem revealed by meta- and proteo-genomics", *ISME Journal*, no. 5, pp. 1735–1747, 2011.

[BID 14] BILD A.H., CHANG J.T., JOHNSON W.E. *et al.*, "A field guide to genomics research", *PLoS Biology*, no. 12, p. e1001744, 2014.

[BOR 15] BORK P., BOWLER C., DE VARGAS C. *et al.*, "Tara Oceans. Tara Oceans studies plankton at planetary scale. Introduction", *Science*, no. 348, p. 873, 2015.

[BOU 11a] BOURDON J., EVEILLARD D., SIEGEL A., "Integrating quantitative knowledge into a qualitative gene regulatory network", *PLoS Computational Biology*, no. 7, p. e1002157, 2011

[BOU 11b] BOUSKILL N.J., EVEILLARD D., O'MULLAN G. *et al.*, "Seasonal and annual reoccurrence in betaproteobacterial ammonia-oxidizing bacterial population structure", *Environmental Microbiology*, no. 13, pp. 872–886, 2011.

[BOW 08] BOWLER C., ALLEN A.E., BADGER J. H. *et al.* "The *Phaeodactylum* genome reveals the evolutionary history of diatom genomes", *Nature*, no. 456, pp. 239–244, 2008.

[BUR 15] BURRI R., NATER A., KAWAKAMI T. *et al.*, "Linked selection and recombination rate variation drive the evolution of the genomic landscape of differentiation across the speciation continuum of *Ficedula* flycatchers", *Genome Research*, no. 25, pp. 1656–1665, 2015.

[CAM 14] CAMPILLO T., RENOUD S., KERZAON I. *et al.*, "Analysis of hydroxycinnamic acid degradation in *Agrobacterium fabrum* reveals a coenzyme A-dependent, beta-oxidative deacetylation pathway", *Applied Environmental Microbiology*, no. 80, pp. 3341–3349, 2014.

[CAR 15] CARRO L., PUJIC P., ALLOISIO N. *et al.*, "*Alnus* peptides modify membrane porosity and induce the release of nitrogen-rich metabolites from nitrogen-fixing *Frankia*", *ISME Journal*, no. 9, pp. 1723–1733, 2015.

[CHA 15] CHAMPOMIER-VERGES M.C., ZAGOREC M. *La métagénomique: Développements et Futures Applications*, Editions Quae, 2015.

[CHA 16] CHASSAING O., DESSE-BERSET N., HÄNNI C. *et al.*, "Phylogeography of the European sturgeon (*Acipenser sturio*): a critically endangered species", *Molecular Phylogenetics and Evolution*, no. 94, pp. 346–357, 2016.

[CHI 13] CHIVIAN D., DEHAL P.S., KELLER K. *et al.*, "MetaMicrobesOnline: phylogenomic analysis of microbial communities", *Nucleic Acids Research*, no. 41, pp. D648–654, 2013.

[COH 06] COHAN F. M., "Towards a conceptual and operational union of bacterial systematics, ecology, and evolution", *Philosophical Transactions of the Royal Society of London B: Biological Sciences*, no. 361, pp. 1985–1996, 2006.

[CON 13] CONDAMINE F.L., ROLLAND J., MORLON H., "Macroevolutionary perspectives to environmental change", *Ecology Letters*, no. 16 Suppl 1, pp. 72–85, 2013.

[COS 13] COSTELLO M.J., MAY R.M., STORK N.E., "Can we name Earth's species before they go extinct?", *Science*, no. 339, pp. 413–416, 2013.

[CRI 14] CRISTESCU M.E., "From barcoding single individuals to metabarcoding biological communities: towards an integrative approach to the study of global biodiversity", *Trends in Ecology and Evolution*, no. 29, pp. 566–571, 2014.

[DAM 11] DAMON C., VALLON L., ZIMMERMANN S. *et al.*, "A novel fungal family of oligopeptide transporters identified by functional metatranscriptomics of soil eukaryotes", *ISME Journal*, no. 5, pp. 1871–1880, 2011.

[DAT 08] BIG DATA S.I.N. "Community cleverness required (Editorial)", *Nature*, no. 455, p. 1, 2008.

[DAV 12] DAVIES N., FIELD D., THE GENOMIC OBSERVATORIES NETWORK, "Sequencing data: a genomic network to monitor Earth", *Nature*, no. 481, p. 145, 2012.

[DAV 15] DAVEY J.W., CHOUTEAU M., BARKER S.L. *et al.*, "Major improvements to the *Heliconius melpomene* genome assembly used to confirm 10 chromosome fusion events in 6 million years of butterfly evolution", *Genes, Genomes, Genetics (Bethesda)*, no. 6, pp. 695–708, 2016.

[DEL 14] DELSUC F., METCALF J.L., WEGENER PARFREY L. *et al.*, "Convergence of gut microbiomes in myrmecophagous mammals", *Molecular Ecology*, no. 23, pp. 1301–1317, 2014.

[DIL 13] DILLIES M.A., RAU A., AUBERT J. *et al.*, "A comprehensive evaluation of normalization methods for Illumina high-throughput RNA sequencing data analysis", *Brief Bioinformatics*, no. 14, pp. 671–683, 2013.

[DUF 08] DUFRESNE A., OSTROWSKI M., SCANLAN D.J. *et al.*, "Unraveling the genomic mosaic of a ubiquitous genus of marine cyanobacteria", *Genome Biology*, no. 9, p. R90, 2008.

[ENA 10] ENARD D., DEPAULIS F., ROEST CROLLIUS H., "Human and non-human primate genomes share hotspots of positive selection", *PLoS Genetics*, no. 6, p. e1000840, 2010.

[FAR 13] FARIELLO M.I., BOITARD S., NAYA H. *et al.*, "Detecting signatures of selection through haplotype differentiation among hierarchically structured populations", *Genetics*, no. 193, pp. 929–941, 2013.

[FAU 15] FAURE D., JOLY D., "Next-generation sequencing as a powerful motor for advances in the biological and environmental sciences", *Genetica*, no. 143, pp. 129–132, 2015.

[FLE 95] FLEISCHMANN R.D., ADAMS M.D., WHITE O. *et al.*, "Whole-genome random sequencing and assembly of *Haemophilus influenzae* Rd", *Science*, no.269, pp. 496–512, 1995.

[FIE 09] FIELD D., SANSONE S.A., COLLIS A. *et al.*, "Megascience, Omics data sharing", *Science*, no. 326, pp. 234–236, 2009.

[FLO 12] FLOUDAS D., BINDER M., RILEY R., "The Paleozoic origin of enzymatic lignin decomposition reconstructed from 31 fungal genomes", *Science*, no. 336, pp. 1715–1719, 2012.

[GAT 11] GATHERERS G.H., "Human evolutionary genomics", *Cell*, no. 146, 2011.

[GEI 13] GEIGL E.M.,"Paléogénétique: faire parler l'ADN ancien", *Biofutur*, no. 349, pp. 1–30, 2013.

[GOM 11] GOMPERT Z., BUERKLE C.A., "A hierarchical Bayesian model for next-generation population genomics", *Genetics*, no. 187, pp. 903–917, 2011.

[GRA 08] GRAVELEY B.R. "Molecular biology: power sequencing", *Nature*, no. 453, pp. 1197–1198, 2008.

[GRA 15] GRANQVIST E., SUN J., OP DEN CAMP R. *et al.*, "Bacterial-induced calcium oscillations are common to nitrogen-fixing associations of nodulating legumes and non-legumes", *New Phytologist*, no. 207, pp. 551–558, 2015.

[GUI 14] GUILLOT G., VITALIS R., ROUZIC A. L. *et al.*, "Detecting correlation between allele frequencies and environmental variables as a signature of selection. A fast computational approach for genome-wide studies", *Spatial Statistics*, no. 8, pp. 145–155, 2014.

[HAG 15] HAGELBERG E., HOFREITER M., KEYSER C., "Introduction. Ancient DNA: the first three decades", *Philosophical Transactions of the Royal Society of London B: Biological Sciences*, no. 370, p. 20130371, 2015.

[HAL 12] HALTER D., GOULHEN-CHOLLET F., GALLIEN S. *et al.*, "In situ proteo-metabolomics reveals metabolite secretion by the acid mine drainage bio-indicator, *Euglena mutabilis*", *ISME Journal*, no. 6, pp. 1391–1402, 2012.

[HAR 13] HARDISTY A., ROBERTS D. "A decadal view of biodiversity informatics: challenges and priorities", *BMC Ecology*, no. 13, p. 16, 2013.

[HEL 12] HELICONIUS GENOME CONSORTIUM., "Butterfly genome reveals promiscuous exchange of mimicry adaptations among species", *Nature*, no. 487, pp. 94–98, 2012.

[HIN 13] HINGAMP P., GRIMSLEY N., ACINAS S.G. *et al.*, "Exploring nucleo-cytoplasmic large DNA viruses in Tara Oceans microbial metagenomes", *ISME Journal*, no. 7, pp. 1678–1695, 2013.

[HOC 11] HOCHER V., ALLOISIO N., AUGUY F. *et al.*, "Transcriptomics of actinorhizal symbioses reveals homologs of the whole common symbiotic signaling cascade", *Plant Physiology*, no. 156, pp. 700–711, 2011.

[HOW 08] HOWE D., COSTANZO M., FEY P. *et al.*, "Big data: The future of biocuration", *Nature*, no. 455, pp. 47–50, 2008.

[HUG 13] HUGONI M., TAIB N., DEBROAS D. *et al.*, "Structure of the rare archaeal biosphere and seasonal dynamics of active ecotypes in surface coastal waters", *Proceedings of the National Academy of Sciences*, no. 110, pp. 6004–6009, 2013.

[IVE 12] IVERSON V., MORRIS R.M., FRAZAR C.D. *et al.*, "Untangling genomes from metagenomes: revealing an uncultured class of marine Euryarchaeota", *Science*, no. 335, pp. 587–590, 2012.

[JOB 12] JOBARD M., RASCONI S., SOLINHAC L. *et al.*, "Molecular and morphological diversity of fungi and the associated functions in three European nearby lakes", *Environmental Microbiology*, no. 14, pp. 2480–2494, 2012.

[JOB 13] JOBLING M., HOLLOX E., HURLES M. *et al.*, *Human Evolutionary Genetics*, GS Garland Science, Taylor & Francis Group., 2013.

[JOL 15] JOLY D., FAURE D., "Next-generation sequencing propels environmental genomics to the front line of research", *Heredity*, no. 114, pp. 429–430, 2015.

[JOR 06] JORON M., JIGGINS C. D., PAPANICOLAOU A. *et al.*, "*Heliconius* wing patterns: an evo-devo model for understanding phenotypic diversity", *Heredity (Edinb)*, no. 97, pp. 157–167, 2006.

[KAR 11] KARSENTI E., ACINAS S.G., BORK P. *et al.*, "A holistic approach to marine eco-systems biology", *PLoS Biology*, no. 9, p. e1001177, 2011.

[KAY 09] KAYE J., HEENEY C., HAWKINS N. *et al.*. "Data sharing in genomics–re-shaping scientific practice", *Nature Review Genetics*, no. 10, pp. 331–335, 2009.

[KLI 11] KLITGORD N., SEGRE D., "Ecosystems biology of microbial metabolism", *Current opinion in biotechnology*, no. 22, pp. 541–546, 2011.

[KRA 10] KRAMER U., "Metal hyperaccumulation in plants", *Annual Review of Plant Biology*, no. 61, pp. 517–534, 2010.

[KRU 11] KRUGER R.P., "Human evolutionary genomics", *Cell*, no. 146, pp. 847–849, 2011.

[LAR 12] LARSON G., KARLSSON E. K., PERRI A. *et al.*, "Rethinking dog domestication by integrating genetics, archeology, and biogeography", *Proceedings of the National Academy of Sciences USA*, no. 109, pp. 8878–8883, 2012.

[LAS 11] LASSALLE F., CAMPILLO T., VIAL L. *et al.*, "Genomic species are ecological species as revealed by comparative genomics in *Agrobacterium tumefaciens*", *Genome Biology and Evolution*, no. 3, pp. 762–781, 2011.

[LAS 15] LASSALLE F., MULLER D., NESME X., "Ecological speciation in bacteria: reverse ecology approaches reveal the adaptive part of bacterial cladogenesis", *Research Microbiology*, no. 166, pp. 729–741, 2015.

[LEH 13] LEHEMBRE F., DOILLON D., DAVID E. *et al.,* "Soil metatranscriptomics for mining eukaryotic heavy metal resistance genes." *Environmental Microbiology*, no. 15, pp. 2829–2840, 2013.

[LI 13] LI Y. F., COSTELLO J.C., HOLLOWAY A.K. *et al.*, "Reverse ecology and the power of population genomics", *Evolution*, no. 62, pp. 2984–2994, 2008.

[LOG 12] LOGARES R., HAVERKAMP T.H., KUMAR S. *et al.*, "Environmental microbiology through the lens of high-throughput DNA sequencing: synopsis of current platforms and bioinformatics approaches", *Journal of Microbiology Methods*, no. 91, pp. 106–113, 2012.

[LOM 12] LOMAN N.J., MISRA R.V., DALLMAN T.J. *et al.*, "Performance comparison of benchtop high-throughput sequencing platforms", *Nature Biotechnology*, no. 30, pp. 434–439, 2012.

[LOS 13] LOSOS J.B., ARNOLD S.J., BEJERANO G. *et al.*, "Evolutionary biology for the 21st century", *PLoS Biology*, no. 11, p. e1001466, 2013.

[MAM 13] MAMINDY-PAJANY Y., HUREL C. *et al.*, "Arsenic in marine sediments from French Mediterranean ports: geochemical partitioning, bioavailability and ecotoxicology", *Chemosphere*, no. 90, pp. 2730–2736, 2013.

[MER 14] MERHEB M., VAIEDELICH S., MANINGUET T. *et al.*, "Molecular species identification in processed animal hides for biodiversity protection", *Int l Journal of Advances in Chemical Engeniery and Biological Sciences (IJACEBS)*, no. 1, pp. 55–57, 2014.

[MER 16] MERHEB M., VAIEDELICH S., MANIGUET T. *et al.*, "Mitochondrial DNA, restoring Beethovens music", *Mitochondrial DNA*, no. 27, pp. 355–359, 2016.

[MON 12] MONCHY S., GRATTEPANCHE J.D., BRETON E. *et al.*, "Microplanktonic community structure in a coastal system relative to a *Phaeocystis* bloom inferred from morphological and tag pyrosequencing methods", *PLoS One*, no. 7, p. e39924, 2012.

[MOR 11] MORA C., TITTENSOR D.P., ADL S., *et al.*, "How many species are there on Earth and in the ocean?", *PLOS Biology*, no. 9, p. e1001127, 2011.

[OGR 87] OGRAM A., SAYLER G.S., BARKAY T. "The extraction and purification of microbial DNA from sediments", *Journal of Microbiological Methods*, no. 7, pp. 57–66, 1987.

[OLL 13] OLLIVIER M., TRESSET A., HITTE C. *et al.*, "Evidence of coat color variation sheds new light on ancient canids", *PLoS One.*, no. 8, p. e75110, 2013.

[ORL 06] ORLANDO L., DARLU P., TOUSSAINT M. *et al.*, "Revisiting Neandertal diversity with a 100,000 year old mtDNA sequence", *Current Biology*, no. 16, pp. R400–402, 2006.

[ORL 15] ORLANDO L., GILBERT M.T., WILLERSLEV E., "Reconstructing ancient genomes and epigenomes", *Nature Review Genetics*, no. 16, pp. 395–408, 2015.

[PAN 15] PANTE E., ABDELKRIM J., VIRICEL A. *et al.*, "Use of RAD sequencing for delimiting species", *Heredity*, no. 114, pp. 450–459, 2015.

[PAP 15] PAPADOPOULOU A., TABERLET P., ZINGER L. "Metagenome skimming for phylogenetic community ecology: a new era in biodiversity research", *Molecular Ecology*, no. 24, pp. 3515–3517, 2015.

[PIO 10] PIONNIER-CAPITAN M., La domestication du chien en Eurasie: étude de la diversité passée, approches ostéoarchéologiques, morphométriques et paléogénétiques, PhD Thesis, École normale supérieure, Lyon, 2010.

[PIT 14] PITTERA J., HUMILY F., THOREL M. *et al.*, "Connecting thermal physiology and latitudinal niche partitioning in marine *Synechococcus*" *ISME Journal*, no. 8, pp. 1221–1236, 2014.

[PLE 13] PLEWNIAK F., KOECHLER S., NAVET B. *et al.*, "Metagenomic insights into microbial metabolism affecting arsenic dispersion in Mediterranean marine sediments", *Molecular Ecology*, no. 22, pp. 4870–4883, 2013.

[PUI 12] PUILLANDRE N., BOUCHET P., BOISSELIER-DUBAYLE M. *et al.*, "New taxonomy and old collections: integrating DNA barcoding into the collection curation process", *Molecular Ecology Resources*, no. 12, pp. 396–402, 2012.

[RAN 13] RANJARD L., DEQUIEDT S., CHEMIDLIN PREVOST-BOURE N. *et al.*, "Turnover of soil bacterial diversity driven by wide-scale environmental heterogeneity", *Nature Communications*, no. 4, p. 1434, 2013.

[RAY 14] RAYNAUD X., NUNAN N., "Spatial ecology of bacteria at the microscale in soil", *PLoS One.*, no. 9, p. e87217, 2014.

[ROE 07] ROESCH L.F.W., FULTHORPE R.R., RIVA A. *et al.*, "Pyrosequencing enumerates and contrasts soil microbial diversity", *ISME Journal*, no. 1, pp. 283–290, 2007.

[ROQ 13] ROQUET C., THUILLER W., LAVERGNE S., "Building megaphylogenies for macroecology: taking up the challenge", *Ecography*, no. 36, pp. 013–026, 2013.

[SAN 92] SANGER F., NICKLEN S., COULSON A.R., "DNA sequencing with chain-terminating inhibitors, 1977", *Biotechnology*, no. 24, pp. 104–108, 1992.

[SAN 12] SANSONE S.A., ROCCA-SERRA P., FIELD D., "Toward interoperable bioscience data", *Nature Genetics*, no. 44, pp. 121–126, 2012.

[SCH 11] SCHLOSS P.D., WESTCOTT S.L., "Assessing and improving methods used in operational taxonomic unit-based approaches for 16S rRNA gene sequence analysis", *Applied Environmental Microbiology*, no. 77, pp. 3219–3226, 2011.

[SCH 14] SCHLOTTERER C., TOBLER R., KOFLER R. *et al.*, "Sequencing pools of individuals – mining genome-wide polymorphism data without big funding", *Nature Review Genetics*, no. 15, pp. 749–763, 2014.

[SHA 13] SHAMS M., VIAL L., CHAPULLIOT D. *et al.*, "Rapid and accurate species and genomic species identification and exhaustive population diversity assessment of *Agrobacterium spp.* using recA-based PCR", *Systematic Applied Microbiology*, no. 36, pp. 351–358, 2013.

[SIM 11] SIME-NGANDO T., NIQUIL N. (eds.), *Disregarded Microbial Diversity and Ecological Potentials Inaquatic Systems*, Springer, The Netherlands, 2011.

[SMA 12] SMADJA CM., CANBÄCK B., VITALIS R. *et al.*, "Large-scale candidate gene scan reveals the role of chemoreceptor genes in host plant specialization and speciation in the pea aphid", *Evolution*, no. 66, pp. 2723–2738, 2012.

[SMA 15] SMADJA CM., LOIRE E., CAMINADE P. *et al.*, "Seeking signatures of reinforcement at the genetic level: a hitchhiking mapping and candidate gene approach in the house mouse", *Molecular Ecology*, no.24, pp. 4222–4237, 2015.

[SOU 14] SOUSA V., PEISCHL S., EXCOFFIER L., "Impact of range expansions on current human genomic diversity", *Current Opinion in Genetics and Development*, no. 29C, pp. 22–30, 2014.

[STA 02] STACKEBRANDT E., FREDERIKSEN W., GARRITY G.M. *et al.*, "Report of the ad hoc committee for the re-evaluation of the species definition in bacteriology", *International Journal of Systematic and Evolutionary Microbiology*, no. 52, pp. 1043–1047, 2002.

[STA 15] STANKOWSKI S., STREISFELD M.A., "Introgressive hybridization facilitates adaptive divergence in a recent radiation of monkeyflowers", *Proceedings of Biological Science*, no. 282, p. 20151666, 2015.

[STE 15] STEPHENS Z.D., LEE S.Y., FAGHRI F. *et al.*, "Big Data: astronomical or genomical?", *PLoS Biol*, no. 13, p. e1002195, 2015.

[SUN 11] SUN S., CHEN J., LI W. *et al.*, "Community cyberinfrastructure for Advanced Microbial Ecology Research and Analysis: the CAMERA resource", *Nucleic Acids Research*, no. 39, pp. D546–551, 2011.

[TAB 12a] TABERLET P., COISSAC E., HAJIBABAEI M. *et al.*, "Environmental DNA", *Molecular Ecology*, no. 21, pp. 1789–1793, 2012.

[TAB 12b] TABERLET P., COISSAC E., POMPANON F. *et al.*, "Towards next-generation biodiversity assessment using DNA metabarcoding", *Molecular Ecology*, no. 21, pp. 2045–2050, 2012.

[TEL 05] TELETCHEA F., MAUDET C., HÄNNI C., "Food and forensic molecular identification: update and challenges", *Trends in Biotechnology*, no. 23, pp. 359–366, 2005.

[TEN 14] TENENBAUM J.D., SANSONE S.A., HAENDEL M., "A sea of standards for omics data: sink or swim?", *The Journal of the American Medical Informatics Association*, no. 21, pp. 200–203, 2014.

[THO 13] THOMAS A.F., RAYMOND M., *"Santé, médecine et sciences de l'évolution: une introduction"*, De Boeck, 2013.

[THO 15] THOMPSON C.C., AMARAL G.R., CAMPEAO M. *et al.*, "Microbial taxonomy in the post-genomic era: rebuilding from scratch?", *Archives of Microbiology*, no. 197, pp. 359–370, 2015.

[VAL 09] VALENTINI A., POMPANON F., TABERLET P., "DNA barcoding for ecologists", *Trends in Ecology and Evolution*, no. 24, pp. 110–117, 2009.

[VAN 12] VAN STRAALEN N.M., ROELOFS D., *An Introduction to Ecological Genomic*, Oxford University Press, Oxford, 2012.

[WAN 15] WANG Y., NAVIN N.E., "Advances and applications of single-cell sequencing technologies", *Molecular Cell*, no. 58, pp. 598–609, 2015.

[WIL 15] WILTING A., COURTIOL A., CHRISTIANSEN P. *et al.*, "Planning tiger recovery: understanding intraspecific variation for effective conservation", *Science Advances*, no. 1, p. e1400175, 2015.

[ZEN 12] ZENGLER K., PALSSON B.O., "A road map for the development of community systems (CoSy) biology", *Nature Review and Microbiology*, no. 10, pp. 366–372, 2012.

[ZHA 14] ZHANG R., MURAT F., PONT C. *et al.*, "Paleo-evolutionary plasticity of plant disease resistance genes", *BMC Genomics*, no. 15, p. 187, 2014.

List of Websites

Chapter 1

1000 Fungal Genomes: http://1000.fungalgenomes.org/home/about/

BarCode of Life: http://www.barcodeoflife.org/

BGI-Europe (Beijing Genomics Insitute): http://bgitechsolutions.com/

DataOne: http://www.dataone.org

Earth Microbiome Project: http://www.earthmicrobiome.org/

e-ReColNat: http://recolnat.org/

Ecoinformatics: http://www.ecoinformatics.org/

Ecological Data: http://ecologicaldata.org

ENCODE: http://www.genome.gov/encode/

G10K: https://genome10k.soe.ucsc.edu/

GO (Genomics Observatories): http://genomicobservatories.blogspot.fr/

Idealg: http://www.idealg.ueb.eu/

i5k: http://www.arthropodgenomes.org/wiki/i5K

IHMC (International Human Microbiome Consortium): http://www.human-microbiome.org/index.php?id=35

JGI (Joint Genome Institute): http://jgi.doe.gov/

MetaGenoPolis: http://www.mgps.eu/index.php?id=homepage

MetaHIT: http://www.metahit.eu/

MycoCosm: http://genome.jgi.doe.gov/programs/fungi/index.jsf

NIEHS: http://www.niehs.nih.gov/

NHGRI: http://www.genome.gov/

NIH: http://www.nih.gov/

Tara Oceans and Oceanomics: http://oceans.taraexpeditions.org/

Terragenome: http://www.terragenome.org/

Chapter 2

454: http://www.454.com/

BGI: http://www.genomics.cn/en/index

France génomique: https://www.france-genomique.org

Illumina: http://www.illumina.com/

Ion Torrent: http://www.iontorrent.com/

JGI: http://www.jgi.doe.gov/

Pacific Biosciences: http://www.pacificbiosciences.com/

Chapter 3

GSC: http://www.standardsingenomics.org/index.php/sigen

EBI: http://www.ebi.ac.uk/

e-infrastructure: http://www.e-irg.eu/

ELIXIR: http://www.elixir-europe.org

ENA: http://www.ebi.ac.uk/ena

DDBJ: http://www.ddbj.nig.ac.jp/intro-e.html

Galaxy: http://galaxyproject.org/

GenBank: http://www.ncbi.nlm.nih.gov/genbank

Gene Expression Omnibus: http://www.ncbi.nlm.nih.gov/geo/

GitHub: https://github.com/

GO: http://genomicobservatories.blogspot.co.uk/

GRISBI: http://www.min2rien.fr/grisbi-%E2%80%93-calcul-scientifique-sur-grille-pour-la-bioinformatique/

IFB: http://www.france-bioinformatique.fr/

Irods: http://irods.org/

MIxS: http://gensc.org/projects/mixs-gsc-project/

NCBI: http://www.ncbi.nlm.nih.gov/

NIG: http://www.nig.ac.jp/english/index.html

R: http://www.r-project.org/

SourceForge:http://sourceforge.net/

Chapter 4

e-Infrastructures pour la Génomique et la Biologie à Grande Echelle: http://www.france-grilles.fr/IMG/pdf/E-InfraGenoBio-rapport_final.pdf

Greengenes: http://greengenes.lbl.gov/

RDP: http://rdp.cme.msu.edu/

SILVA: http://www.arb-silva.de/

Pplacer: http://matsen.fhcrc.org/pplacer/

MEGAN: http://ab.inf.uni-tuebingen.de/software/megan5/

GSC:http://gensc.org/

Barcode of Life: http://www.barcodeoflife.org/

Galaxy: http://galaxyproject.org/

BioConductor de R:https://www.bioconductor.org/ ; https://cran.r-project.org/

Chapter 5

ITIS: http://www.itis.gov/

Chapter 6

BoL:http://www.barcodeoflife.org/

Silva:http://www.arb-silva.de/

RDP: http://rdp.cme.msu.edu/

PR2: http://ssu-rrna.org/

Chapter 8

www.eva.mpg.de/neandertal

Chapter 11

Galaxy: http://galaxy.prabi.fr/

LTER: http://www.lter-europe.net/

Chapter 12

Future Earth: http://www.futureearth.org/

Index

Printed in the United States
By Bookmasters